Lecture Notes in Computer Science

Edited by G. Goos and J. Hartmanis

277

B. Benninghofen
S. Kemmerich
M. M. Richter

Systems of Reductions

Springer-Verlag
Berlin Heidelberg New York London Paris Tokyo

Authors

Benjamin Benninghofen
MBB
D-8012 Ottobrunn, FRG

Susanne Kemmerich
Technische Hochschule Aachen
Lehrstuhl für Angewandte Mathematik, insbesondere Informatik
Templergraben 64, D-5100 Aachen, FRG

Michael M. Richter
Universität Kaiserslautern, Fachbereich Informatik
Postfach 30 49, D-6750 Kaiserslautern, FRG

CR Subject Classification (1987): F.2.2, I.2.3, F.4.1, F.4.2

ISBN 3-540-18598-4 Springer-Verlag Berlin Heidelberg New York
ISBN 0-387-18598-4 Springer-Verlag New York Berlin Heidelberg

2145/3140-543210

Introduction

Systems of reductions (or rewrite rules as they are often called) enjoy a growing popularity in theoretical computer science. They have also become a useful tool in computational algebra; these areas are anyway not very well separated and have much in common. An important branch of this common background is "equational logic". This is, strictly speaking, the fragment of predicate logic, where equality is the only predicate. In practice, equality logic is concerned with a class of problems which are more restricted as well as more general. Typical are implications of the form

$$\Sigma \Rightarrow P.$$

Here Σ is a universally quantified set of equations; P, however, need not to be an equation and not even a formula of first order predicate logic.

An example of the latter type of problems is:

"Is each algebra in which Σ holds finite?"

or equivalently

"Is the free algebra defined by Σ finite?"

The answer one expects in equality logic to such questions is an <u>algorithmic</u> one. That means, one does not consider single questions but whole classes of questions. These may arise in different ways:

1) Σ is fixed and P varies.
   An example is the word problem, where Σ defines the algebra and P varies over all equations $s = t$.

2) Σ varies and P is fixed.
   An example is the infinity problem for a class of algebras (which are defined by the axiom system Σ under consideration).

3) Σ and P both vary.
   An example is the uniform word problem for a class of algebras.

Algorithmic decision problems arise in computer science mainly in the context of abstract data types and automated theorem proving. The algorithmic tradition in mathematics, in particular in algebra, is much older. A hundred years ago, at the time of Kronecker, solutions to algebraic problems were more or less automatically expected to be computable. In the first half of the twentieth century abstract and non-constructive methods became more dominant. There was still an enormous interest in principal aspects of computability and decision problems, but constructive methods did not play such a big role in the every-day-work of mathematicians. This has radically changed under the influence of the development of computer science. Today constructive methods are not only regarded as being useful in certain applied situations; their main purpose is to provide relevant structural and combinatorial insights.

Among the defining properties of equality reflexivity is the only trivial one. The basic idea of a reduction is to give up symmetry and allow only replacements of equals by equals in one direction. This idea is as old as the word problem for groups which was considered by M. Dehn and A.Thue in 1912-14 and which was one of the first decision problems to be formulated. In fact, Dehn's algorithm is the application of a specific system of reductions and its properties were studied for more than fifty years. Giving up symmetry looses, of course, some of the power of equality. The research on Dehn's algorithm was concerned with the question which aspects of full equality are preserved by its one-sided use.The idea of regaining part of the power provided by symmetry by systematically adding new reductions came up much later. This leads to concepts of systems of reductions like the Church-Rosser and the weak Church-Rosser property (which are known under various names) and the finite termination property; in connection with the latter there is growing interest in partial and total well-founded orderings on the terms.

The most useful property is certainly completeness, it ensures that each term t reduces to a uniquely determined irreducible term $t^{\#}$; $t^{\#}$ is the canonical form of t. The aim of the completion algorithm is to enlarge a given system of reductions in order to obtain a complete one. A complete system (if it exists) can be regarded as a link between the finite system of equations and the algebra defined by these equations which is a set-theoretic, often infinite, object.

There are two main lines of research here. On the one hand one studies the completion algorithm and searches for criteria which ensure its termination. As the completion algorithm in many (one is tempted to say "most") cases fails to terminate this leads to the investigation of infinite systems. In many cases these can be finitely described and are as useful as finite systems.

The other type of investigations is concerned with the use of complete systems. A complete system certainly provides an answer to the word problem but unravels much more of the structure of the algebra under investigation. This turns out to be most apparent in the case of groups.

Most of the material in these notes was obtained in the years after 1978 at the Technische Hochschule Aachen; it is partially contained in the dissertations of Hans Bücken, Patrick Horster and Susanne Kemmerich; Patrick Horster also wrote section IV.2.Very useful for computer experiments was an implementation of the completion algorithm for groups and semigroups as well as the forward-backward algorithm. This implementation was done by Klaus Dittrich in Pascal on a Cyber 175. The last part of these notes was written by Friedrich Otto; the material is part of his Habilitationsschrift at the University of Kaiserslautern.

The main interest of the authors is in general principles. Most concrete applications are in group and semigroup theory, however. There are several reasons for this. One is that these are familiar structures and one has a better feeling for difficulties and importance of results than in general universal algebras. Another reason is that in these areas computational methods are well established. This gives possibilities for interesting connections and comparisons and is also useful for a fruitful competition.

In order to make the volume somewhat self-contained much general material is included. The idea was to provide the reader with an at least almost complete introduction. Here completeness is meant in the sense that suffices for an understanding of the rest of the material. It is natural that many aspects had to be left out.

The authors have also been very reluctant with historical remarks. On the one hand, many results seem to be obtained independently by different authors. On the other hand, the situation was not so clear to us that we dared to make statements on priority questions.

There are several people whom the authors are indebted for useful help and discussions over the years. Among the former students of Aachen we would name Hans Bücken, Klaus Dittrich, Petra Zimmermann and Tom Beske. One of the authors wants to mention Dallas Lankford in particular; he was also early influenced by Woody Bledsoe. Later on useful discussions took place with Mike Ballantyne, Bruno Buchberger, Richard Göbel, Deepak Kapur, Wolfgang Küchlin and J. Neubüser.

All authors are indebted to Mr. v. Hehl and Mrs. Scarlet Nökel for the excellent preparation of the manuscript.

Last not least our thanks are due to the Deutsche Forschungsgemeinschaft without their support over the years this research would not have been possible.

Contents

I. General Concepts from Universal Algebra

I.1. Algebras, Terms and Substitutions

Although we assume familiarity with the basic concepts of algebra we recall some of the definitions. We will only be concerned with finitary algebras, i.e. algebras of the form

$$\mathbf{A} = \langle A, f_i^A \mid 1 \leq i \leq m \rangle$$

where the $f_i^A\colon A^{n_i} \to A$ are n_i-ary operations, $n_i \geq 0$ a natural number. If $n_i=0$ we call f_i^A a constant and identify it with an element from A, the carrier of **A**.
The signature of **A** is the sequence $\nu = \nu(\mathbf{A}) = \langle n_i \mid 1 \leq i \leq m \rangle$; we fix it throughout most of the discussion.

Homomorphisms $h\colon \mathbf{A} \to \mathbf{B}$ are structure preserving mappings from A to B, i.e. $h(f_i^A(a_1,\ldots,a_{ni})) = f_i^B(h(a_1),\ldots,h(a_{ni}))$ holds.
Isomorphisms are 1-1 homomorphisms and endomorphisms are homomorphisms $\mathbf{A} \to \mathbf{A}$.
$\mathbf{A} \cong \mathbf{B}$ means that **A** and **B** are isomorphic.

A congruence relation for **A** is an equivalence relation $R \subseteq A^2$ such that for $1 \leq k \leq n_i$

$$R(a_k, b_k) \quad \text{implies} \quad R(f_i^A(a_1,\ldots,a_{ni}), f_i^A(b_1,\ldots,b_{ni})).$$

The congruence class of $a \in A$ modulo R is denoted by $[a]_R$ or simply by [a]; the quotient of A modulo R is

$$^{A}/_{R} = \{[a] \mid a \in A\};$$

the canonical mapping $\pi: A \to {}^{A}/_{R}$ is given by $\pi(a) = [a]$. Defining the operations on $^{A}/_{R}$ by

$$f_i^{A/R}([a_1],\dots,[a_{ni}]) = [f_i^{A}(a_1),\dots,f_i^{A}(a_{ni})]$$

one obtains the quotient algebra $^{\mathbf{A}}/_{R}$.

The congruence relation induced by a homomorphism $h: \mathbf{A} \to \mathbf{B}$ is $R_h = \{(a,b) \mid h(a) = h(b)\}$; it is also called the kernel of h.

The basic interrelations between these concepts are:

(i) The canonical mapping is a homomorphism $\mathbf{A} \to {}^{\mathbf{A}}/_{R}$ the kernel of which is R.

(ii) If $h: \mathbf{A} \to \mathbf{B}$ is a surjective homomorphism with kernel R_h, then $^{\mathbf{A}}/R_h \approx \mathbf{B}$.

The intersection of congruence relations is again a congruence relation. Therefore for each relation $R \subseteq A^2$ there is a smallest congruence relation $\langle R \rangle$ s.t. $R \subseteq \langle R \rangle \subseteq A^2$. This is called the congruence generated by R; the operator $R \to \langle R \rangle$ is a closure operator.

A congruence relation R is fully invariant if R(a,b) implies R(h(a),h(b)) for each endomorphism h. Again, for each $R \subseteq A^2$ there is the smallest fully invariant congruence relation $R \subseteq \langle\!\langle R \rangle\!\rangle$; the operator $R \to \langle\!\langle R \rangle\!\rangle$ is a closure operator too.

In some sense the main topic of these notes is the investigation of this operator from the algorithmic point of view.

For a set X an algebra **A** with $X \subseteq A$ is called an <u>absolutely free algebra generated by X</u> iff

> for every algebra $\mathbf{B} = \langle B, f_i^B \mid 1 \le i \le m \rangle$ (of the same signature as **A**) and every mapping $h: X \to B$ there is a unique extension of h to a homomorphism $\underline{h}: \mathbf{A} \to \mathbf{B}$.

Another equivalent possibility of describing an absolutely free algebra **A** is:

(i) If $f_i^A(t_1,\ldots,t_{ni}) = f_j^A(s_1,\ldots,s_{nj})$, then $i=j$, $t_k=s_k$ for all k, $1 \le k \le n_i$.

(ii) $f_i^A(t_1,\ldots,t_{ni}) \notin X$ for all operations f_i^A and all arguments t_k.

(iii) **A** is generated by X, i.e. if **B** is a subalgebra of **A** such that $X \subseteq B$, then **B** = **A**.

An example is the algebra $\langle \mathbf{N}, s \rangle$ of the natural numbers with the successor operation generated by the singleton set {0}. In this case the conditions (i) - (iii) are just the Peano axioms which motivates the alternative name <u>Peano algebra</u> instead of <u>absolutely free algebra</u>.

An absolutely free algebra generated by X is unique up to isomorphism, hence we can refer to the absolutely free algebra **A**(X) generated by X. It always exists and there are various ways of constructing it, e.g. as strings or as trees over an alphabet $X \cup \{f_i \mid 1 \le i \le m\}$ where the f_i are symbols not in X.

For absolutely free algebras **A** the <u>principle of structural induction</u> holds:

> If a set $S \subseteq A$ contains all generators and if $t_1,\ldots,t_{ni} \in S$ implies $f_i^A(t_1,\ldots,t_{ni}) \in S$ for all operations f_i^A and all terms t_k, then S = A.

In particular two special cases are of interest:

a) $X = \emptyset$:
A($\emptyset$) is called the <u>initial algebra</u> or the <u>algebra of words</u> and is denoted by **W**. (It is only of interest if there are constants because it would be empty otherwise).

b) $X = Var = \{x_0, x_1, \ldots\}$ is a countably infinite set:
In this case **T** = **A**(Var) is called the <u>algebra of terms</u> (with the carrier T) or the <u>term algebra</u>; the $x \in Var$ are the <u>variables</u>.

When talking about initial algebras we make the general assumption that we have constants. The explicitly constructions allow us to regard terms as strings or as trees. The principle of structural induction gives the possibility to define familiar concepts like "s is a subterm of t" or "a variable x occurs in t"; we can also refer to particular occurrences of s in t.
We can regard the algebra of words as a subalgebra of the term algebra; words are also called <u>ground terms</u>. While the homomorphisms **T** → **A** are in one-one correspondence with the mappings Var → A there is exactly one homomorphism h: **W** → **A** for each **A**. In particular, each algebra generated by the empty set is a quotient of the word algebra. We also note that the identity is the only endomorphism of the word algebra.

The concept of an absolutely free algebra refers to the class of all algebras.
If **K** is any class of algebras then **A** is called <u>free for **K**</u> over X iff

> **A** ∈ **K**, $X \subseteq A$ and for any **B** ∈ **K**, **B** = $\langle B, f_i^B \mid 1 \le i \le m \rangle$ and for every mapping h: X → B there is a unique extension of h to a homomorphism $\underline{h}$: **A** → **B**.

For arbitrary classes $\mathbf{K}$ free algebras will not always exist. In connection with equational theories there are however sufficiently many free algebras. Again it is easy to see that two free algebras for $\mathbf{K}$ over X are isomorphic.

An <u>equation</u> is an ordered pair $\langle s,t \rangle$ of terms and will be denoted by $s \equiv t$; a <u>ground equation</u> is an equation between ground terms.

A homomorphism $h: \mathbf{T} \to \mathbf{A}$ <u>satisfies</u> an equation $s \equiv t$ iff $h(s) = h(t)$; $s \equiv t$ is <u>true</u> in $\mathbf{A}$ iff every homomorphism $h: \mathbf{T} \to \mathbf{A}$ satisfies $s \equiv t$. This last situation is denoted by $\mathbf{A} \Vdash s \equiv t$.

The <u>models</u> of a set E of equations are those algebras for which $\mathbf{A} \Vdash s \equiv t$ holds for all $s \equiv t \in E$ (denoted by $\mathbf{A} \Vdash E$).

The <u>semantic consequences</u> of a set $\mathbf{E}$ of equations are those equations $s \equiv t$ for which $\mathbf{A} \Vdash E$ implies $\mathbf{A} \Vdash s \equiv t$; the notation for this is $E \Vdash s \equiv t$.

<u>Theories</u> are sets of equations closed under consequences. A theory E <u>defines a class</u> $\mathbf{K}$ of algebras iff $\mathbf{K} = \{\mathbf{A} \mid \mathbf{A} \Vdash E\}$.

A class $\mathbf{K}$ of algebras is <u>equationally definable</u> (or a <u>variety</u>) iff $\mathbf{K}$ is defined by some theory E; in this case the class is denoted by $\mathbf{K}(E)$.

If $\langle E \rangle \supseteq E$ is the congruence relation generated by E in $\mathbf{T}$ then the quotient $\mathbf{T}_E = \mathbf{T}/_{\langle E \rangle}$ is generated by $X_E = \pi_E(X)$ where $\pi_E: \mathbf{T} \to \mathbf{T}/_{\langle E \rangle}$ is the canonical mapping. It is important to notice:

$$\mathbf{T}_E \text{ is free for } \mathbf{K}(E) \text{ over } X_E.$$

Hence varieties have free algebras; moreover we mention that a class of algebras has free algebras iff it is a variety.

In the same way $\mathbf{W}_E = \mathbf{W}/_{\langle E\rangle}$ is an initial algebra in $\mathbf{K}(E)$ if E contains only ground equations. The equations true in $\mathbf{K}(E)$ are the same as the equations true in $\mathbf{T}_E$ and these are in general more than just $\langle E\rangle$.
In fact $\{s \equiv t \mid \mathbf{T}_E \Vdash s \equiv t\} =: \langle\!\langle E\rangle\!\rangle$ is the fully invariant congruence relation generated by E.
Because the word algebra has no non-trivial endomorphism we have $\langle\!\langle E\rangle\!\rangle = \langle E\rangle$ in the ground case.

Because initial algebras are unique up to isomorphisms we can refer to $\mathbf{W}_E$ as the initial algebra of $\mathbf{K}(E)$. Such initial algebras are also called <u>abstract</u> <u>data</u> <u>types</u>, E is called a <u>specification</u> of the abstract data type. Here we recall that we assume a fixed signature which in principle has to be added to the specification.
The initial algebra $\mathbf{W}_E$ contains all information of equality logic of $\mathbf{K}(E)$ and this information is again contained in the congruence relation $\langle E\rangle$ generated by E. In fact, $\langle E\rangle$ contains just the semantic consequences of E:

$$\langle E\rangle = \{s \equiv t \mid E \Vdash s \equiv t\}.$$

$\langle\!\langle E\rangle\!\rangle$ has been defined as the smallest fully invariant congruence relation above E. A construction of $\langle\!\langle E\rangle\!\rangle$ from "below" is as follows. It works for arbitrary algebras $\mathbf{A}$ but we are only interested in the cases $\mathbf{A} = \mathbf{T}$ and $\mathbf{A} = \mathbf{W}$.
For $D \subseteq A \times A$ we put

(i) $R(D) = D \cup \{\langle t, t\rangle \mid t \in A\}$

(ii) $E(D) = D \cup \{\langle h(s),h(t)\rangle \mid \langle s,t\rangle \in D$ and h is an endomorphism of $\mathbf{A}\}$

(iii) $S(D) = D \cup \{\langle t, s\rangle \mid \langle s, t\rangle \in D\}$

(iv) $C(D) = D \cup \{\langle f_i^A(t_1,\dots,t_{ni}), f_i^A(s_1,\dots,s_{ni})\rangle \mid \langle t_k,s_k\rangle$ from D for $1 \le k \le n_i$, $1 \le i \le m\}$

(v) $T(D) = D \cup \{(s,t) \mid \exists\, t_1,\dots,t_n,\ t_1=s,\ t_n=t,$ $\langle t_k,t_{k+1}\rangle \in D$ for $1 \le k < n\}$.

R(D), S(D) and T(D) are the closure operations under reflexivity, symmetry and transitivity; C(D) guarantees a congruence relation and E(D) makes it fully invariant.
In this terminology the generated congruence relation is

$$\langle\langle E\rangle\rangle = T(C(S(E(R(E)))))$$

which is straightforward to show.

The conditions for the formation of the sets E(D), S(D), C(D) and T(D) can in an obvious way be viewed as rules for their generation. That means, $\langle\langle E\rangle\rangle$ can be obtained from E and the reflexive equations $t \equiv t$ by iterated applications of these rules. In particular $\langle\langle E\rangle\rangle$ is recursively enumerable whenever E is recursively enumerable.
The closure under endomorphisms will be considered below when we discuss substitions. It can be omitted in two important cases:

a) In the ground case where we have $\langle\langle E\rangle\rangle = \langle E\rangle$ because the word algebra has no non-trivial endomorphisms;
b) If E is closed under endomorphisms which occurs often in applications.

In order to illustrate the terminology we consider the familiar examples of (finitely presented) monoids and groups. Here we have a binary operation "$\cdot$", a constant e, a constant a_k for each generator, $1 \le k \le m$ and an additional unary operation "$^{-1}$" in the group case.
If the only equations are $x\cdot(y\cdot z) = (x\cdot y)\cdot z$, $e\cdot x = x$ (and $x^{-1}\cdot x = e$ resp.) then the word algebra $\mathbf{W}_E$ is the <u>free monoid</u> (resp. the <u>free group</u>) on m generators, denoted by $FM(a_1,\dots,a_m)$ (resp. by $FG(a_1,\dots,a_m)$).
If the presentation contains additional ground equations $S = \{s_i \equiv t_i \mid 1 \le i \le n\}$, then $FM(a_1, \dots, a_m)/\langle S\rangle$ and $FG(a_1,\dots,a_m)/\langle S\rangle$ are called the monoid resp. the group defined by that presentation.

In the case of groups the equations can be written in the form $u_i = s_i \cdot (t_i)^{-1} \equiv e$, hence the presentation is given by the words u_i, $1 \le i \le n$.
The quotient of $FG(a_1,\dots,a_m)$ modulo $\langle S \rangle$ is then the quotient $FG(a_1,\dots,a_m)/N$ where N is the normal subgroup generated by the u_i, $1 \le i \le n$.
The elements of FM are congruence classes and each such class contains a unique word u which is either e or of the form $(b_1 \cdot (b_2 \cdot (\dots)\dots)$ where $b_i \neq e$ and b_i is a generator; in the group case we have in addition $b_i \neq {b_{i+1}}^{-1}$. These terms are written as $b_1 \cdot b_2 \dots b_k$; they are usually called <u>semigroup</u> resp. <u>group</u> <u>words</u> or simply <u>words</u>.
As usual we will identify the carrier of FM resp. FG with the set of these words; the identity element e is also denoted by 1.

<u>Substitutions</u> are endomorphisms $\theta: T \to T$ of the term algebra such that $\theta(x) = x$ for all except finitely many variables x. A substitution is thus determined by all values $\theta(x) = s$ for which $s \neq x$, x a variable, and will in general be defined in this way. The set of substitutions is closed under composition.
For a term t and a substitution θ the term $\theta(t)$ is called an <u>instance</u> of t.
If $\theta(x)$ is always a variable and θ is one-one then θ is called a <u>renaming</u> <u>of</u> <u>variables</u>.
If V_1 and V_2 are two finite sets of variables then there are always two renaming substitutions θ_1 and θ_2 such that $\theta_1(V_1) \cap \theta_2(V_2) = \emptyset$; in this case we say that θ_1 and θ_2 <u>separate</u> the variables of V_1 and V_2.

The notion of substitution does not cover the concept of replacement of certain subterms of a term by other terms. We need the following terminology:

1. Definition:

(i) A place is a tuple $\lambda = (i_1,\ldots,i_n)$ of natural numbers; the empty place ( ) is also admitted.

(ii) The subterm of a term t at place λ is inductively defined by

$$sbt(t,(\ )) = t$$
$$sbt(f(t_1,\ldots,t_n),(i,\lambda)) = sbt(t_i,\lambda) \quad \text{for } 1 \le i \le n$$
and undefined otherwise.

(iii) The replacement of the subterm occuring at a place λ in a term t by a term r is inductively defined by

$$rep(t,(\ ),r) = r$$
$$rep(f(t_1,\ldots,t_n),(i,\lambda),r) = f(t_1,\ldots,rep(t_i,\lambda,r),\ t_{i+1},\ldots,t_n)$$
for $1 \le i \le n$ and undefined otherwise.

2. Definition:

(i) A unifier of a set S of terms is a substitution θ such that $\theta(s) = \theta(t)$ for all $s,t \in S$.
S is called unifiable iff it has a unifier.

(ii) A most general unifier of S (m.g.u.(S) in short) is a unifier θ of S such that for all unifiers σ of S there is some substitution λ satisfying $\sigma = \lambda \cdot \theta$:

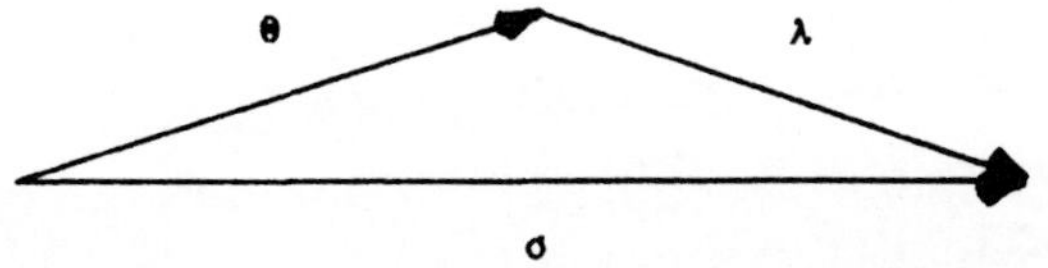

If S = {s,t} then we indicate the most general unifier σ of S by σ = m.g.u.(s,t). This functional notion is justified by the fact that most general unifiers are unique up to a renaming of variables.

We quote the fundamental theorem about most general unifiers.

3. The Unification Theorem:

Let S be a finite set of terms.

(i) It is decidable whether S is unifiable.

(ii) S is unifiable iff it has a most general unifier.

(iii) Most general unifiers can be constructed effectively.

In fact, there is an algorithm which decides the existence of unifiers and constructs a most general one in the positive case, cf. [Ro 65].

Subsitutions are special cases of more general replacements called reductions. The latter are the main topic of this volume and will be introduced in chapter II.1..
Reductions provide a basis for computations in term algebras. Such computations are closely connected with ordering relations on the terms.

4. Definition:

Suppose $\prec$ is an irreflexive and transitive relation over the term algebra T.

(i) $\prec$ is called a <u>term ordering</u> iff for all terms s,t,r, $s \prec t$ and all substitutions θ we have $\theta(s) \prec \theta(t)$ and $rep(r,\lambda,s) \prec r$ if $sbt(r,\lambda) = t$.

(ii) $\prec$ is called <u>well-founded</u> iff there are no infinite chains $t_1 \succ t_2 \succ \ldots$.

(iii) $\prec$ is called <u>total</u> iff for all $s \neq t$ either $s \prec t$ or $t \prec s$ holds.

In a term ordering no two variables x,y can be comparable because $x \prec y$ implies by the substitution property $y \prec x$; transitivity gives $x \prec x$, a contradiction to irreflexivity.

In particular, no term ordering can be total.

On the other hand, in many cases the restriction of the term ordering to the ground terms is total.
An important class of such orderings will be widely used in the sequel; it was introduced in [Kn-Be].

5. Definition:

Let $\mathbf{T} = \langle T, f_i \mid 1 \le i \le m \rangle$ be a term algebra.
A weight function ω for $\mathbf{T}$ is a function

$$\omega: \{f_i \mid 1 \le i \le m\} \cup Var \rightarrow \mathbb{N} \quad \text{such that}$$

(i) $\omega(f_i) > 0$ for $n_i=1$, $1 \le i < m$ and for $n_i=0$, $1 \le i \le m$

(ii) $\omega(x) = \min(\omega(f_i) \mid n_i=0)$ for all $x \in Var$.

The weight function is extended to T by

$$\omega(s) = \sum_{x \in Var} n(x,s) \cdot \omega(x) + \sum_{i=1}^{m} n(f_i,s) \cdot \omega(f_i)$$

where $n(x,s)$ (resp. $n(f_i,s)$) denotes the number of occurences of x in s (resp. of f_i in s).

The so called KB-orderings (which will be defined next) depend on these weight functions.

6. Definition:

The KB-ordering $\prec = \prec(\omega)$ on a term algebra T associated with a weight function ω is defined recursively by:

$s \prec t$ iff

(i) $\omega(s) < \omega(t)$ and $n(x,s) \leq n(x,t)$ for all $x \in Var$,

or (ii) $\omega(s) = \omega(t)$ and $n(x,s) = n(x,t)$ for all $x \in Var$ and either

(α) $s = x$ and $t = f_m(f_m(\ldots f_m(x)\ldots)$ for some $x \in Var$

or (β) $s = f_k(s_1,\ldots,s_{nk})$, $t = f_j(t_1,\ldots,t_{nj})$ such that (a) $k < j$

or (b) $k = j$ and for some i, $1 \leq i \leq n_k$, $s_l = t_l$ for all $l < i$ and $s_i \prec t_i$.

The recursive call is in the last clause of this definition.

By direct inspection we get immediately that each KB-ordering is a term ordering.

7. Proposition:

(i) The KB-orderings are well-founded.

(ii) The restriction of a KB-ordering to the ground terms is total.

For the proofs we refer to [Kn-Be]; they are lengthy but do not contain any surprises.

Additional remarks:

There have been various attempts to built in additional axioms **E** into the substitutions. This leads to a modification of the unification problem:

Are two terms s and t unifiable modulo **E** ?

An important case is where **E** consists of the equations of associativity and commutativity which was carried out in [Sti 81]. A comprehensive overview of unification theory is given in [Si 84].

Besides the KB-orderings more general orderings have been investigated. One way for a generalization is not to require totality even for the ground terms (such a partial ordering will occur in chapter IV). Well-founded total orderings on the ground terms are isomorphic (as ordered structures) to ordinal numbers. The order type of all KB-orderings on ground terms is a very simple one, namely that of the natural numbers. There have been used orderings of a much bigger order type, however, e.g. the recursive path orderings of Dershowitz (cf. e.g. [Der 82] or [Der 85]).

I.2 Some Concepts of the Theory of Formal Languages and Automata

In this paragraph we will review some of the basic concepts of the theory of formal languages and automata, which will be used in several of the following chapters. The main purpose of this is setting the terminology and listing the tools and principles of automata theory needed furtheron in order to make the book self contained. A detailed development of the material of this section occurs in any introductory text to the theory of formal languages and automata (cf. [Ho-Ul], [Ha 78]). The results about derivatives of regular expressions can be found in [Br 64].

A finite alphabet $\{a_1,\ldots,a_n\}$ will usually be denoted by Σ, while Σ^* is the set of all words $w = b_1\ldots b_m$, $b_i \in \Sigma$ for $1 \leq i \leq m$. The length of a word w, denoted by $|w|$, is the number of symbols of w. The word of length 0 is the empty word, mostly denoted by e or λ. Notice that we always have $e \in \Sigma^*$. Hence we denote by $\Sigma^+ := \Sigma^* \setminus \{e\}$ the set of all non-empty words over Σ.

A finite automaton (or nondeterministic finite state acceptor) A is a 5-tuple (Q,Σ,δ,q_o,F), where Q is a finite set of states, Σ is a finite input alphabet, $q_o \in Q$ is the initial state, $F \subseteq Q$ is the set of final states and δ is the transition function mapping $Q \times \Sigma$ to the power set $P(Q)$. Thus a finite automaton is a mathematical model of a system with discrete inputs and a finite control (the set of states Q) which determines the behaviour of the system on subsequent inputs.

A finite automaton is often described by its <u>transition diagram</u>, a labelled directed graph, where the vertices correspond to the states of the finite automaton and an arrow from state q_i to state q_j labelled with a is contained in the transition diagram iff $q_j \in \delta(q_i,a)$.
If the range of δ contains only one-element sets, i.e. if δ maps the set $Q \times \Sigma$ into Q, A is called <u>deterministic</u>.
The transition function δ can be extended to Σ^* in the following way:

$$\delta(q_i,wa) := \{ q' \mid \exists q \in \delta(q_i,w) \text{ such that } q' \in \delta(q,a) \}$$

for $w \in \Sigma^*$ and $a \in \Sigma$.

A word $w \in \Sigma^*$ is said to be <u>accepted</u> by the finite automaton A iff $\delta(q_o,w) \cap F \neq \emptyset$. The set of all words, i.e. the <u>language</u>, accepted by a finite automaton A is denoted by <u>L(A)</u>. A set of words $T \subseteq \Sigma^*$ is called <u>regular</u> if there is some finite automaton A such that T = L(A).

A wider class of languages can be recognized by the so called <u>pushdown automata</u> which include besides to the finite control a more powerful memory, the <u>stack</u>. A pushdown automaton B is a 7-tuple $(Q,\Sigma,\Gamma,\delta,q_o,Z_o,F)$, where Q is a finite set of <u>states</u>, Σ is a finite <u>input alphabet</u>, Γ is a finite <u>stack alphabet</u>, $q_o \in Q$ is the <u>initial state</u>, $Z_o \in \Gamma$ is the <u>start symbol</u>, $F \subseteq Q$ is the set of <u>final states</u> and δ is the <u>transition function</u> mapping $Q \times (\Sigma \cup \{e\}) \times \Gamma$ to finite subsets of $Q \times \Gamma^*$.
The language accepted by a pushdown automaton B is the set of all input words for which some choice of moves causes B to enter a final state; it will be denoted by <u>L(B)</u> and is called <u>context-free</u>. Again context-free sets are exactly those which can be accepted by some pushdown automata.

The Turing machines, at least, accept just the recursive enumerable sets. A Turing machine has a finite control (i.e. a finite set of states Q, with $q_o \in Q$ the start state and $F \subseteq Q$ the set of final states), an infinite input tape, that is divided into cells containing exactly one symbol and a tape head which scans one cell at one time. The tape is input- as well as output medium; finitely many connected cells of it contain symbols of Σ, the input alphabet, while the remaining infinitely many cells are inscribed with a special symbol not in Σ, the symbol blank B. A Turing machine is started at the leftmost input symbol in state q_o; its possible moves are in dependency of the state $q \in Q$ and the input symbol $a \in \Sigma \cup \{B\}$, scanned by the tape head, the following:

- replace a by $a' \in \Sigma \cup \{B\}$ and change the state without moving the tape head
- move the tape head one cell to the right or to the left without changing the tape and change the state

Formally a Turing machine TM is a 6-tuple $(Q,\Sigma,\delta,q_o,B,F)$, where Q,Σ,q_o,B,F are as desribed above and δ is the next move function mapping $Q \times (\Sigma \cup \{B\})$ to $Q \times (\Sigma \cup \{B\}) \times \{left, right\}$.
A configuration of a Turing machine TM is denoted by $\alpha_1 q \alpha_2$, where q is the current state of TM and $\alpha_1\alpha_2$ is the string in $(\Sigma \cup \{B\})^*$ that is the contents of the tape from the leftmost non-blank symbol up to the rightmost non-blank symbol or the symbol left to the tape head; we assume that the tape head scans the leftmost symbol of α_2.
If $a_1 \ldots a_{i-1} q a_i \ldots a_n$ is a configuration and $\delta(q,a_i)=(p,a_j,left)$ resp. $\delta(q,a_i)=(p,a_j,right)$ we write:

$$a_1 \ldots a_{i-1} q a_i \ldots a_n \vdash_{TM} a_1 \ldots a_{i-2} p a_{i-1} a_j a_{i+1} \ldots a_n$$

$$\text{resp. } a_1 \ldots a_{i-1} q a_i \ldots a_n \vdash_{TM} a_1 \ldots a_{i-1} a_j p a_{i+1} \ldots\ldots a_n$$

and say that the second configuration results from the first by one move.

The language <u>L(TM)</u> accepted by a TM = $(Q,\Sigma,\delta,q_o,B,F)$ is:

$$L(TM) := \{ w \mid w\in\Sigma^* \text{ and } q_o w \vdash^*_{TM} \alpha_1 p \alpha_2 \text{ for some } p\in F \text{ and } \alpha_1\alpha_2 \in (\Sigma\cup\{B\})^* \}$$

here $\vdash^*_{TM}$ is the reflexive, transitive closure of $\vdash_{TM}$.

Besides the characterization of different kinds of languages by accepting machines one can define these languages with the aid of generation processes, the so called grammars.
A <u>grammar</u> is a 4-tuple G = (N,T,s,P), where N (the set of <u>non-terminals</u>) and T (the set of <u>terminals</u>) are disjoint alphabets, $s\in N$ is called <u>start symbol</u> and P is a finite set of ordered pairs (u,v) of words over $N\cup T$, the elements of $P \subseteq V^* \times V^*$, $V:=N\cup T$, are called <u>productions</u> and will be denoted by $u \to v$, while $\to^*$ is the reflexive and transitive closure of $\to$.

The language <u>L(G)</u> generated by a grammar G is defined by

$$L(G) := \{ w \mid w\in T^* \text{ and } s \to^* w \}$$

According to the kind of productions allowed, the Chomsky hierarchy distinguishes four classes of grammars (the so called <u>type-i grammars</u> i=0,..,3) in the following way:

type-0: (unrestricted grammars)
$u \to v \in P$ implies $u \in V^*$, $v \in V^*$

type-1: (context-sensitive grammars)
$u \to v \in P$ implies $|u| \leq |v|$

type-2: (context-free grammars)
$u \to v \in P$ implies $u \in N$, $v \in V^+$

type-3: (regular grammars)

$u \rightarrow v \in P$ implies $u \in N,\ v \in T^*$

A language is called <u>type-i language</u> and is denoted by L_i, $i=0,..,3$ if it is generated by a type-i grammar.
A well-known theorem of the theory of formal languages is:

$$L_3 \subset L_2 \subset L_1 \subset L_0$$

Furthermore the following equivalences hold:

type-0 grammars are equivalent to Turing machines
type-2 grammars are equivalent to pushdown automata
type-3 grammars are equivalent to finite automata

i.e.

L is of type-0 iff L is recursive enumerable
L is of type-2 iff L is context-free
L is of type-3 iff L is regular

For regular sets there is even a third way of characterisation which is based upon the fact that the class of regular sets is closed under union, concatenation and Kleene closure. (Notice that the class of regular sets is also closed under intersection and complement).

Thus a regular set can be defined as the set of those elements which can be achieved by repeatedly applying the operations '$\cup$', '*' and '$\cdot$' to elements of Σ; each operation may be applied only finitely many times.

Furthermore, we can describe regular sets by simple expressions, the so called <u>regular expressions</u>, which are defined as follows:

(i) $\emptyset$, e and a for every $a \in \Sigma$ are regular expressions denoting $\emptyset$, {e} and {a}, respectively.

(ii) If S,T are regular expressions denoting regular sets $\mathcal{S}$ and $\mathcal{T}$, then (S+T), ST and S^* are regular expressions denoting $\mathcal{S} \cup \mathcal{T}$, $\mathcal{S}\mathcal{T} := \{ st \mid s \in \mathcal{S}, t \in \mathcal{T} \}$ and $\mathcal{S}^* := \cup(\mathcal{S}^i / i \in \mathbb{N})$, where $\mathcal{S}^o := \{e\}$ and $\mathcal{S}^i := \mathcal{S}\mathcal{S}^{i-1}$ for $i \geq 1$.

For formal manipulations with regular expressions are the notions of derivatives (cf. [Br 64]) useful.
The <u>left-derivative</u> <u>${}_wD(T)$</u> of a regular expression $T \subseteq \Sigma^*$ according to a word $w \in \Sigma^*$ is the following set:

$$ {}_wD(T) := \{ t \mid wt \in T \} $$

Analogously one obtains the <u>right-derivative</u> <u>$D_w(T)$</u> by:

$$ D_w(T) := \{ t \mid tw \in T \} $$

Notice that ${}_wD(T)$ and $D_w(T)$ are regular and that there is an algorithm which generates a regular expression ${}_wD(T)$ resp. $D_w(T)$ for every regular expression T and every word w. Furthermore, there are only finitely many different derivatives for a regular expression T.
Let w be a word over Σ^* with the property that for all $w' \in \{ v \mid {}_vD(T) = {}_wD(T) \}$ (resp. for all $w' \in \{ v \mid D_v(T) = D_w(T) \}$) we have $w < w'$, with respect to some ordering '<' over Σ^*, then w is called a <u>left-characteristic</u> <u>word</u> of the regular expression T (resp. a <u>right-characteristic</u> <u>word</u> of T). Of course, there are only finitely many characteristic words for every regular expression T.
By <u>$CH_l(T)$</u> resp. <u>$CH_r(T)$</u> we denote the set of all left-characteristic resp. right-characteristic words belonging to T.

Further we need in chapter III.1 the following terminology and lemmas:

1. Definition:

Let $T \subseteq \Sigma^*$ be a regular expression and $l \in \Sigma^*$ some left-characteristic and $r \in \Sigma^*$ some right-characteristic word of T.

We call the sets:

$$_{l}S(T) := \{ s \mid {_s}D(T) = {_l}D(T) \} \cap l\Sigma^*$$
$$S_r(T) := \{ s \mid D_s(T) = D_r(T) \} \cap \Sigma^* r$$

the <u>left-supplement of T</u> according to l resp. the <u>right-supplement of T</u> according to r.

2. Lemma:

The left-supplement $_{l}S(T)$ and the right-supplement $S_r(T)$ are regular for every regular expression T and every left-characteristic word l resp. every right-characteristic word r of T.

Proof: The proposition results from

$$_{l}S(T) = \bigcup_{r \in {_l}D(T)} D_r(T) \cap t\Sigma^*$$

and the closure property.

[]

3. Lemma:

Let T,L and I be regular expressions with the property that for all $l \in L$ $l \notin \Sigma^* L \setminus \{l,e\} \Sigma^*$.

(i) $T \subseteq IL$ yields $T = \sum_{i=1}^{n} I_i L_i$, where $I_i \in I, L_i \in L$ for $1 \le i \le n$

(ii) $T \subseteq LI$ yields $T = \sum_{i=1}^{n} L_i I_i$, where $L_i \in L, I_i \in I$ for $1 \le i \le n$

Proof: (i) For $\{l_1, \ldots l_n\} := CH_r(T) \cap L$ we have $T = \sum_{i=1}^{n} D_{l_i}(T) l_i$

Hence $T = \sum_{i=1}^{n} D_{l_i}(T)(S_{l_i}(T) \cap L)$ because for every $r \in D_{l_i}(T)$ we have $rs \in T$ for all $s \in L$ with $D_s(T) = D_{l_i}(T)$.

Thus we have to show: $I_i := D_{l_i}(T) \subseteq I$

Assume there is some $r \in D_{l_i}(T)$ such that $r \notin I$.

Since $rl_i \in IL$ we have either:

(a) $l_i = l_1 l_2$ with $l_1 \neq e$, $rl_1 \in I$ and $l_2 \in L$ and therefore $l_i \in \Sigma^* L \setminus \{l_i, e\}$ which is impossible by the assumption of the lemma.

(b) $r = r_1 r_2$, $r_2 \neq e$, $r_1 \in I$ and $r_2 l_i \in L$ and therefore $r_2 l_i \in \Sigma^* L \setminus \{r_2 l_i, e\}$ which is again impossible.

(ii) Analogously to (i) one obtains:

$$T = \sum_{i=1}^{n} l_i \; {}_{l_i}D(T) = \sum_{i=1}^{n} ({}_{l_i}S(T) \cap L)\, {}_{l_i}D(T) = \sum_{i=1}^{n} L_i I_i$$

with $L_i = {}_{l_i}S(T) \cap L \subseteq L$, $I_i = {}_{l_i}D(T) \subseteq I$ for $\{l_1, \ldots, l_n\} := CH_l(T) \cap L$.

[]

Obviously the following lemma holds:

4. Lemma:

Let P, Q, P_i, Q_i be regular expressions with $P_i \subset P$, $Q_i \subset Q$ for $1 \le i \le n$. Then

$$PQ \setminus \sum_{i=1}^{n} P_i Q_i = (P \setminus \sum_{i=1}^{n} P_i)Q + \sum_{i=1}^{n} P_i(Q \setminus Q_i).$$

5. Definition:

For a regular expression T we call:

(i) $\alpha(T) := \{ D_{l_i}(T) \mid l_i \in CH_r(T) \}$
the *set of initial parts of T*.

(ii) $\varepsilon(T) := \{ {}_{l_i}D(T) \mid l_i \in CH_l(T) \}$
the *set of end parts of T*.

Obviously $\alpha(T)$ and $\varepsilon(T)$ are regular for regular T.

Finally, let us recall the *generalized sequential machines*, *GSM* for short, and the *GSM mappings* which describe our regular reduction systems in chapter III.1..
A GSM is a finite automaton with output; formally a GSM is a 6-tuple $\sigma = (Q, I, O, \delta, q_o, F)$, where Q is a finite set of *states*, I the *input-* and O the *output alphabet*, q_o is the *initial state*, $F \subseteq Q$ is the set of *final states* and δ is a mapping from $Q \times I$ to the finite subsets of $Q \times O^*$, where $\delta(q,a) = (p,w)$ means that σ in state q with input symbol a may (as a possible choice of move) enter state p and emit the word w.

The domain of δ can be extended to $Q \times I^*$ in the following way:

$$\delta(q,e) = \{(q,e)\}$$
$$\delta(q,wa) = \{(p,v) \mid v=v_1v_2 \text{ and for some } p' \ (p',v_1)\in\delta(q,w) \text{ and } (p,v_2)\in\delta(p',a)\}$$

Furthermore, the <u>GSM mapping</u> σ determined by some GSM σ is defined as follows:

$$\sigma(w) := \{ v \mid (p,v) \in \delta(q_0,w) \text{ for some } p\in F \} \text{ ; thus}$$
$$\sigma(T) := \{ v \mid v \in \sigma(t) \text{ for } t \in T \}.$$

The GSM mapping is called deterministic if the corresponding GSM is deterministic.
For a GSM mapping σ: $P(I^*) \rightarrow P(O^*)$ and regular $T \subseteq I^*$ resp. $S \subseteq O^*$ the sets $\sigma(T)$ resp. $\sigma^{-1}(S)$ are always regular.

I.3 Decidability

With each term algebra and each set **E** of equations a number of decision problems is associated. In general most of these problems turn out to be recursively unsolvable even in the case of finitely presented groups. We will shortly review these results which form the background for these notes.

Suppose **A** is the free algebra definded by **E**.

The word problem

For arbitrary terms s and t: Does $\mathbf{A} \Vdash s \equiv t$ hold ?

This is equivalent to: Does $\mathbf{E} \Vdash s \equiv t$ hold ?

E. Post and A. Markov proved independently the unsolvability of the word problem for semigroups, cf. [Dav 58]. Examples of groups with an unsolvable word problem were first given by W.W. Boone (a structured proof due to L.J. Britton is given in [Rot 73]) and P.S. Novikov (cf. [Nov 58]).
A consequence of the Higman, Neumann, Neumann embedding theorem (in [HNN 49]) is that there are such groups even with two generators.

The word problem is in some sense the father of the algebraic decision problems. It is, however, only concerned with the elements of the algebra. For applications in computer science this suffices in many cases. One is e.g. often interested in the individual behaviour of certain programs but not so much in properties of whole classes of programs (although this may be disputed).

For algebraists in general algebraic properties of **A** are of more importance.

Among the undecidable properties for groups we find the following:

- Is **A** finite ?
- Is **A** cyclic ?
- Is **A** commutative ?
- Is **A** simple ?
- Is **A** solvable ? (cf. [Ad 55] or [Rab 58]).

It is natural to ask for restricted situations where such questions have a positive answer. Two classical examples of this kind are:

- The word problem for abilian groups [Rei 50]
- The word problem for one relator groups [Mag 32].

One purpose of the methods discussed in these notes is to obtain more cases with positive answers to various decision problems. That such answers cannot be totally uniform even in the group case is shown in [Boo-Ro 66] :

> There is no algorithm which solves the word problem for all groups which have a solvable word problem.

In case where we present undecidability proofs the problems are reduced to:

a) The halting problem for Turing machines

or

b) The intersection problem for type-0-languages (i.e. is $L(G_1) \cap L(G_2) = \emptyset$ for type-0-grammars G_i, where $L(G_i)$ is the language generated by G_i, $i=1,2$)

or

c) The question whether a type-0-grammar allows an infinitely long derivation.

II. Finite Sets of Reductions

II.1. First Concepts

For some set E of equations and some terms t and s the notion $E \Vdash s \equiv t$ of semantical consequence means that $s \equiv t$ is true in all models of E. As indicated in chapter I.1. there are various ways to describe a syntactic consequence operator $\vdash$ with the property $\vdash \; = \; \Vdash$. In particular, the set $\langle E \rangle$ of consequences of E is recursively enumerable as long as E itself is recursively enumerable. On the other hand E is in general not recursive because the word problem is in general undecidable.

In the sequel we will always (tacitly) assume that in a given set E of equations different equations have disjoint sets of variables occuring in them. This can always be achieved by renaming the variables, if necessary.

Informally, a system of reductions is a restricted syntactic operator which does not necessarily prove all consequences of E, but in many cases does nevertheless decide the relation $E \Vdash t \equiv s$. Essentially a system of reductions will be a finite system of equations which can only be applied from the left to the right side. This means, we often loose the symmetry of the equality sign. From the computational point of view this is an advantage, however, since one avoids the loops provided by the symmetric equality sign.

1. Definition:

(i) A system of reductions is a set **R** of ordered pairs ⟨t,s⟩ of terms which are called reductions and will be denoted by $t \rightarrow s$.

(ii) A term v is an immediate reduction of a term u by $t \rightarrow s$ (denoted by $u \;{}_{(t,s)}{\rightarrow}\; v$) iff for some substitution θ the term $\theta(t)$ occurs as a subterm in u and v is the result of replacing some occurence of $\theta(t)$ in u by $\theta(s)$.

(iii) $u \;{}_{\mathbf{R}}{\rightarrow}\; v$ means $u \;{}_{(t,s)}{\rightarrow}\; v$ for some $t \rightarrow s$ in **R**.

(iv) ${}_{\mathbf{R}}{\rightarrow}^{*}$ is the reflexive and transitive hull of ${}_{\mathbf{R}}{\rightarrow}$.

(v) ${}_{\mathbf{R}}{\rightarrow}^{+}$ is the transitive hull of ${}_{\mathbf{R}}{\rightarrow}$.

If no ambiguity arises the reference to **R** is omitted.

A more formal definition of an immediate reduction of u by $t \rightarrow s$ at a place λ (cf. Definition 1 of chapter I.1.) is:

$u \;{}_{(t,s)\lambda}{\rightarrow}\; v$ iff for some substitution θ:

(a) $sbt(u,\lambda) = \theta(t)$

(b) $v = rep(u, \lambda, \theta(s))$.

It is again assumed that no two reductions have a variable in common.

2. Definition:

The semantic consequences of $\{s \equiv t \mid s \rightarrow t \in \mathbf{R}\}$ form the equational theory of **R**.

3. Definition:

(i) **R** and **R'** are called equationally equivalent iff they have the same equational theory.

(ii) **R** and **R'** are derivation equivalent iff $s \xrightarrow[R]{*} t$ implies $s \xrightarrow[R']{*} t$ and vice versa.

Derivation equivalent systems are equationally equivalent but the converse is not true.

Systems of reductions are generalizations of grammars known from the theory of formal languages. Analogous to formal language concepts one defines for a term t:

4. Definition:

(i) $\underline{L(t,\mathbf{R})} = \{s \mid t \xrightarrow[R]{*} s\}$,
$\underline{L^{+}(t,\mathbf{R})} = \{s \mid t \xrightarrow[R]{+} s\}$

(ii) t is called irreducible with respect to **R** iff $L^{+}(t,\mathbf{R}) = \emptyset$; otherwise t is reducible.

(iii) $\underline{Irr} = \underline{Irr(\mathbf{R})} = \{t \mid t \text{ irreducible}\}$
$\underline{Red} = \underline{Red(\mathbf{R})} = \{t \mid t \text{ reducible}\}$

(iv) $\underline{Irred(t,\mathbf{R})} = Irr \cap L(t,\mathbf{R})$
$\underline{Irred(S,\mathbf{R})} = \cup(Irred(t,\mathbf{R}) \mid t \in S)$ for a set S of terms.

Of course, any set of equations may be regarded as a system of reductions. Giving up symmetry this way for the benefit of a better computational behaviour provides immediately two problems:

1) Regarding equations as reductions may not avoid infinitely long chains of derivations or even loops as long as there is no restriction "applying the law of transitivity"; e.g. $s \equiv t$ and $t \equiv s$ may both be reductions.

2) Using the equations as reductions may loose too much information; e.g. there might be no way to find out that a certain equation $s \equiv t$ is a consequence of E.

For this reason some nice properties are introduced which a system of reductions may have.

<u>5. Definition:</u>

R has the <u>finite termination property FTP</u> iff there are no infinite chains $t_n \;{}_{\mathbf{R}}\!\!\rightarrow t_{n+1}$, $t_n \neq t_{n+1}$, $n \in \mathbb{N}$.

<u>6. Definition:</u>

(i) **R** has the <u>Church-Rosser property CR</u> iff for all t and $t_1, t_2 \in L(t,\mathbf{R})$ we have $L(t_1,\mathbf{R}) \cap L(t_2,\mathbf{R}) \neq \emptyset$. This means for $t \;{}_{\mathbf{R}}\!\!\rightarrow^{*} t_1$, $t \;{}_{\mathbf{R}}\!\!\rightarrow^{*} t_2$ there is some s such that $t_1 \;{}_{\mathbf{R}}\!\!\rightarrow^{*} s$ and $t_2 \;{}_{\mathbf{R}}\!\!\rightarrow^{*} s$ hold:

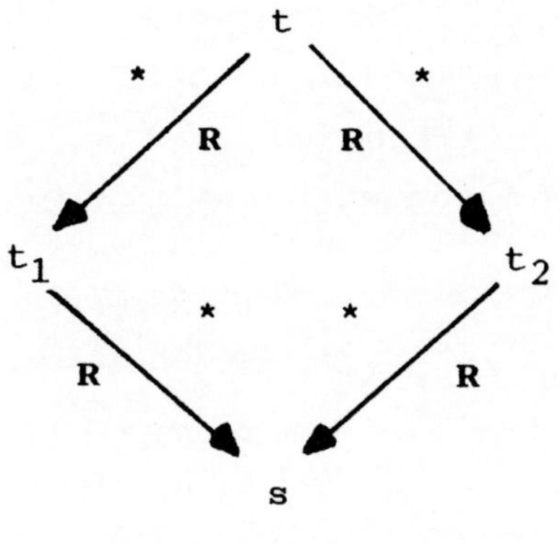

(ii) **R** has the <u>weak</u> <u>Church-Rosser</u> <u>property</u> <u>WCR</u> iff for $t \underset{\mathbf{R}}{\rightarrow} t_1$, $t \underset{\mathbf{R}}{\rightarrow} t_2$ there is some s such that $t_1 \underset{\mathbf{R}}{\overset{*}{\rightarrow}} s$ and $t_2 \underset{\mathbf{R}}{\overset{*}{\rightarrow}} s$ hold:

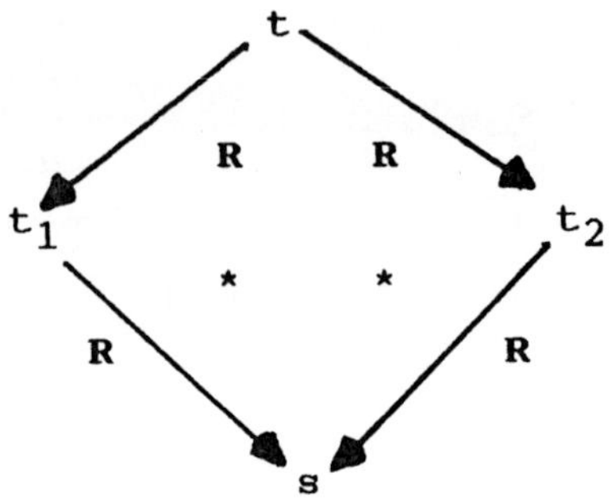

<u>7. Definition:</u>

R has the <u>unique</u> <u>termination</u> <u>property</u> <u>UTP</u> iff **R** has FTP and Irred(t,**R**) contains exactly one term for each t.

It is immediate that the two properties UTP and CR coincide in the presence of FTP.

<u>8. Definition:</u>

R is <u>complete</u> iff **R** has FTP and CR.

The finite termination property FTP means that the relation $s \underset{\mathbf{R}}{\overset{*}{\rightarrow}} t$ is well-founded. As any well-founded relation it also allows an <u>induction</u> <u>principle</u> which we will state here explicitely:

(IP) Suppose P is a unary predicate on the set of terms. If for any t we have: If P(s) for all $s \neq t$ such that $t \underset{\mathbf{R}}{\overset{*}{\rightarrow}} s$, then P(t); then P(t) for all t.

For a complete set of reductions **R** any process of rewriting a term t eventually stops at a unique irreducible term t^*; t^* can be regarded as the <u>normal form</u> of t with respect to **R**.

There are various relations between the concepts introduced above. First we see that even in the ground case WCR does not imply CR.

9. <u>Example:</u>
0,**0**,a,b,c,d are constants and f is a unary function symbol.
R:= {0 → **0**, 0 → a, **0** → b, **0** → c, a → f(a), c → f(a), c → d, d → f(c), d → f(b), b → f(b)}.
WCR is easily checked and the following diagram indicates the absence of CR because L(a,**R**) ∩ L(b,**R**) = ∅.

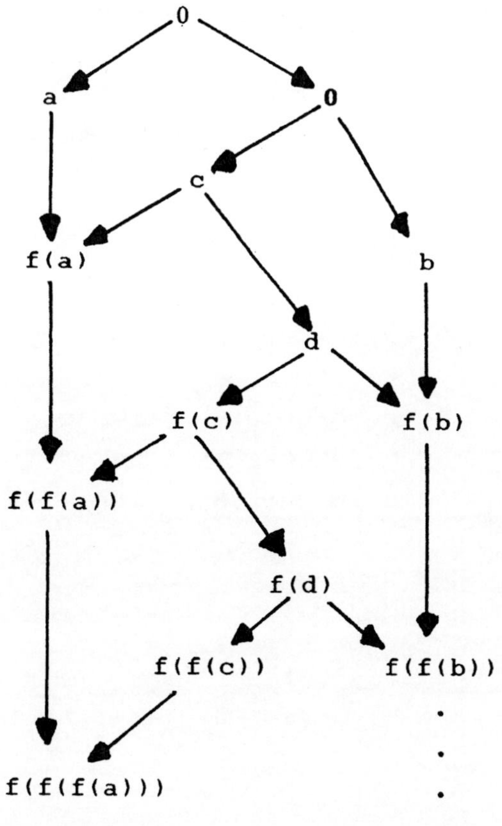

The reason for the failure of CR in the above example is the existence of infinite chains of reductions. This is not an accident as the next proposition shows.

<u>10. Proposition:</u>

Suppose **R** has FTP. Then **R** has WCR iff **R** has CR.

<u>Proof:</u> One implication is trivial.
For the other direction we assume WCR. We consider the following predicate P(x):

$$P(t) \Leftrightarrow [(\forall t_1,t_2 \in L(t,\mathbf{R}))(t \neq t_1, t \neq t_2 \Rightarrow L(t_1,\mathbf{R}) \cap L(t_2,\mathbf{R}) \neq \emptyset)].$$

The induction principle (IP) will show that P(t) holds for all terms t:

Assume $t \ {}_{\mathbf{R}}{\to}^{*}\ s_1$, $t \ {}_{\mathbf{R}}{\to}^{*}\ s_2$, $s_1 \neq t \neq s_2$ and take t_1, t_2 such that $t_1 \neq t \neq t_2$ and $t \to t_1 \to^{*} s_1$, $t \to t_2 \to^{*} s_2$.

By WCR we get $t_1 \to^{*} u$ and $t_2 \to^{*} u$ for some u.

The assumptions $P(t_1)$, $P(t_2)$ and $P(u)$ provide

$$v_1 \in L(s_1,\mathbf{R}) \cap L(u,\mathbf{R}),$$
$$v_2 \in L(s_2,\mathbf{R}) \cap L(u,\mathbf{R})$$

and finally $s \in L(v_1,\mathbf{R}) \cap L(v_2,\mathbf{R})$

such that:

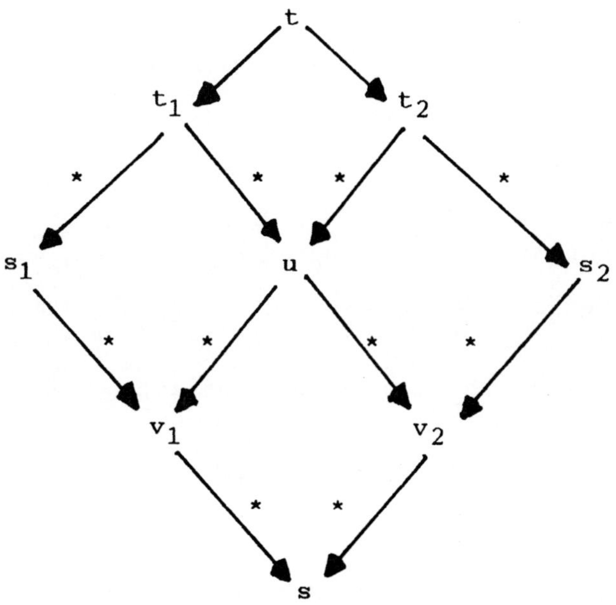

Hence P(t) follows and by (IP) we obtain P(t) for all t. Because only the trivial cases $t \to^* s_1$, $t \to^* s_2$ with $s_1 = t$ or $s_2 = t$ are remaining the assertion CR is proved.

[]

Besides the induction principle (IP) König's lemma is sometimes useful. It says that an infinite finitely branched tree contains an infinite branch. An application is:

11. Proposition:

Assume that **R** has FTP and that for each t there are only finitely many s such that $t \; {}_{\mathbf{R}}\!\!\to s$ holds.
Then each L(t,**R**) is finite.

The proof is immediate by König's lemma. Note that the second assumption is satisfied for finite **R**.

There are various decision problems connected with the properties introduced above. We begin with some undecidability results.

<u>12. Theorem:</u>

Undecidable properties for finite systems of reductions are:

(i) The finite termination property FTP.

(ii) The Church-Rosser property CR.

<u>Proof:</u> The decision problems for FTP and CR will be reduced to undecidable problems for type-0-languages.

Suppose $\mathbf{G} = \langle V_N, V_T, a_o, P\rangle$ is a type-0-grammar where $V_N = \{a_o,\dots,a_n\}$ and $V_T = \{a_{n+1},\dots,a_{n+m}\}$.

We choose some $x \notin V := V_N \cup V_T$ and put $V_x^* = \{wx \mid w \in V^*\}$; the mapping

$$\psi: V^* \to V_x^*, \quad \psi(w) := wx \quad \text{is a bijection.}$$

Let $\mathbf{T}_x = \langle T_x; A_o,\dots,A_{n+m}, f\rangle$ be the absolutely free algebra with n+m+1 constants and one binary operation generated by the singleton {x}. $\mathbf{T}_x$ is a term algebra with one variable.

Next we define two injective mappings

$$\varphi: V^* \to T_x, \quad \varphi_x: V_x^* \to T_x$$

by

$$\varphi(u) = \begin{cases} x & \text{for } u = e \\ A_i & \text{for } u = a_i \\ f(A_i,\varphi(w)) & \text{for } u = a_i w,\ w \neq e \end{cases}$$

and

$$\varphi_x(u) = \begin{cases} x & \text{for } u = x \\ f(A_i, \varphi_x(w)) & \text{for } u = a_i w \end{cases}$$

Finally we put

$$\mathbf{R(G)} = \{\langle\varphi(u),\varphi(v)\rangle, \langle\varphi_x(\psi(u)),\varphi_x(\psi(v))\rangle \mid \langle u,v\rangle \in P\}$$
$$\cup \{\langle f(A_i,\varphi(u)),A_i\rangle, \langle u,e\rangle \in T \mid 0 \le i \le n+m\}$$

and show

$$(*) \quad u \ {}_{P}\!\!\to v \quad \text{iff} \quad \varphi(u) \ {}_{\mathbf{R(G)}}\!\!\to \varphi(v).$$

(a) Assume $u \to v$ by the production rule $\langle r,s\rangle$.
We have the following cases:

(i)	$u = r,$	$v = s$	
(ii)	$u = w_1 r,$	$v = w_1 s$	
(iii)	$u = r w_2,$	$v = s w_2$	
(iv)	$u = w_1 r w_2,$	$v = w_1 s w_2$	where $w_1 \neq e \neq w_2$.

Case (i) is immediate. For case (ii) we have for $s \neq e$

$$\varphi(w_1 r) \ {}_{\langle\varphi(r),\varphi(s)\rangle}\!\!\to \ \varphi(w_1 s)$$

because $\varphi(r)$ is a subterm of $\varphi(w_1 r)$ and $\varphi(s)$ is the corresponding subterm of $\varphi(w_1 s)$; similar for $s = e$.
For case (iii) we consider the substitution $\theta(x) = \varphi(w_2)$ and get

$$\theta(\varphi_x(\psi(r))) = \varphi(r w_2) \quad \text{and} \quad \theta(\varphi_x(\psi(s))) = \varphi(s w_2)$$

which yields

$$\varphi(r w_2) \ {}_{\langle\varphi x(\psi(r)),\varphi x(\psi(s))\rangle}\!\!\to \ \varphi(s w_2).$$

Case (iv) is as (iii).

Therefore in all cases $\varphi(u) \xrightarrow[R(G)]{} \varphi(v)$ holds.

(b) The other direction is an immediate consequence of the injectivity of the mappings ψ, φ and φ_x and the fact that $\varphi(u)$ contains x iff u = e.

From (*) it follows that **G** has infinite derivations iff **R(G)** does not have FTP; hence FTP is undecidable.

If G_1 and G_2 are two grammars (w.l.o.g. with disjoint non-terminals $(a_o,\ldots,a_n,\ a_o',\ldots,a_m')$) then we add a new non-terminal a and productions $a \to a_o$, $a \to a_o'$ to the union of G_1 and G_2.

Then $L(G_1) \cap L(G_2) \neq \emptyset$ iff the following square can be completed.

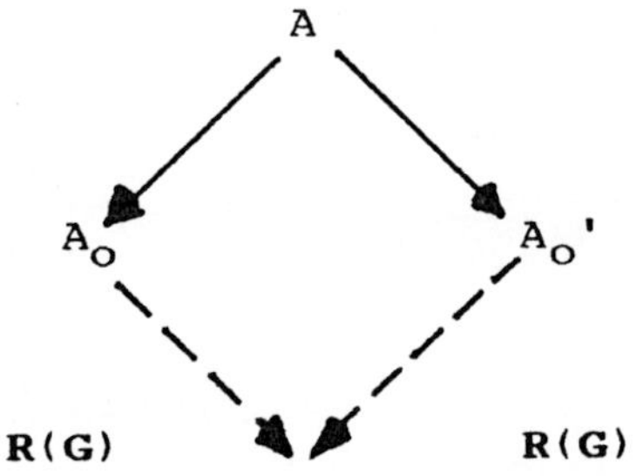

This implies that CR is also undecidable. []

Although FTP and CR are in general undecidable properties one is interested in necessary or sufficient criteria. While CR will be investigated in the next section we will shortly discuss FTP here.

<u>13. Definition:</u>

R is called <u>compatible</u> with some term ordering $\prec$ iff $t \prec s$ for all $s \to t \in$ **R**.

<u>14. Proposition:</u>

(i) If **R** is compatible with some term ordering $\prec$ which is well-founded then **R** has FTP.

(ii) For each system **R** with FTP the relation

$$t \prec s \quad \text{iff} \quad s \ {}_{\mathbf{R}}\!\!\to^{*} t$$

is a well founded term ordering compatible with **R**.

<u>Proof:</u> (i) The conditions of a term ordering ensure $t \prec s$ also for all s,t such that $s \ {}_{\mathbf{R}}\!\!\to^{*} t$.

(ii) If $t \prec s$ is defined as $s \ {}_{\mathbf{R}}\!\!\to^{*} t$ then condition (i)(a) of Definition 4 in chapter I.1. follows from the fact that $t \ {}_{\mathbf{R}}\!\!\to^{*} s$ implies $\theta(t) \ {}_{\mathbf{R}}\!\!\to^{*} \theta(s)$ for all substitutions θ; (i)(b) is immediate.

[]

In general a term ordering will contain more pairs $t \prec s$ than pairs $s \ {}_{\mathbf{R}}\!\!\to^{*} t$ for some system **R** compatible with $\prec$. Term orderings are defined independent of any equational theory. If an equational theory is given, the ideal situation is that each congruence class contains a $\prec$-minimal term and that **R** reduces the terms to the minimal ones. This problem leads to the investigations in chapter II.2..

II.2. The Completion Algorithm

Incompleteness of a system **R** of reductions restricts its applications and can be regarded as some kind of a defect: It does not have enough reductions in order to estimate the power of such a system (and to enlarge it, if wanted and possible) it is important to detect such a defect. The general undecidability results restrict such attempts. As we have seen in chapter II.1. however, FTP reduces CR to WCR. In order to obtain further reductions of WCR to special cases we introduce some terminology.

1. Definition:

(i) Left(**R**) $= \{t \mid t \to s \in \mathbf{R}$, for some $s\}$

Right(**R**) $= \{s \mid t \to s \in \mathbf{R}$, for some $t\}$

(ii) superpos(t_1,t_2) $= \{\sigma(t_1) \mid \sigma$ a m.g.u.(t,t_2) for some subterm t of t_1, $t \notin$ Var$\}$

(iii) If S_1, S_2 are sets of terms then

superpos(S_1,S_2) $= \bigcup(\text{superpos}(t_1,t_2) \mid t_1 \in S_1,\ t_2 \in S_2)$

(iv) If $t_1, t_2 \in$ Left(**R**), $t_1 \to s_1$, $t_2 \to s_2 \in$ **R** and $\sigma(t_1) \in$ superpos(t_1,t_2), then the pair (w_1,w_2) with

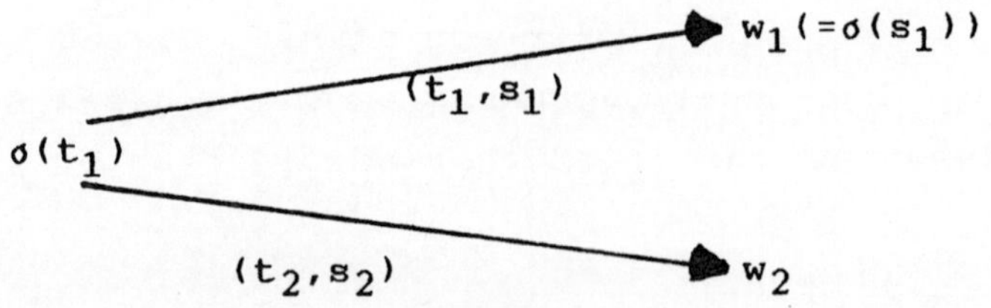

is called a critical pair of **R**.

Note that a finite reduction system **R** has only finitely many critical pairs and each superpos(t_1,t_2) is finite.
The next decidability result is fundamental for many further algorithmic considerations.

2. Theorem:

If **R** is finite and has FTP then the Church-Rosser property CR is decidable.

Proof: It suffices to determine the validity of WCR. We consider a situation of the form

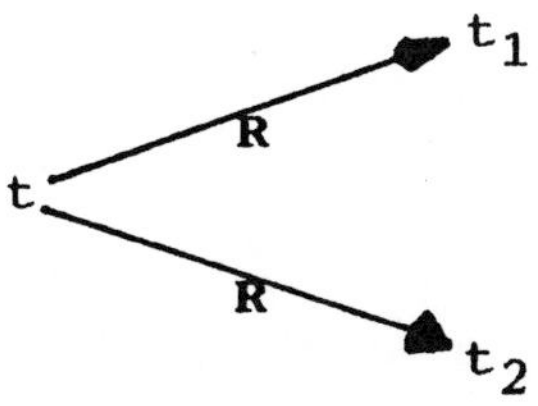

A simple structural induction in the term algebra shows that we can restrict ourselves to the case where $t = \theta(s)$ for some $s \in$ Left(**R**) and θ a substitution.
Finally we will see that we can restrict ourselves to finitely many substitutions θ.
We first remark that $t_1 \;{}_{(t,s)}\!\!\to s_1$ implies $\lambda(t_1) \;{}_{(t,s)}\!\!\to \lambda(s_1)$ for every substitution λ.
Now assume $\theta(t_1) \;{}_{(t,s)}\!\!\to w$ for some $t_1 \to s_1$ in **R** and some substitution θ. Then there is some subterm r of t_1 and some substitution ξ such that $\xi(t) = \theta(r)$ and replacing $\xi(t)$ by $\xi(s)$ at some occurence in $\theta(t_1)$ gives w.
Neglecting some standardizing substitutions which separate variables we can assume $\xi = \theta$ which means that t and r are unifiable and therefore have a m.g.u. $\sigma = \sigma(t,r)$. For some λ then $\theta = \lambda\cdot\sigma$.

We have the commutative diagram

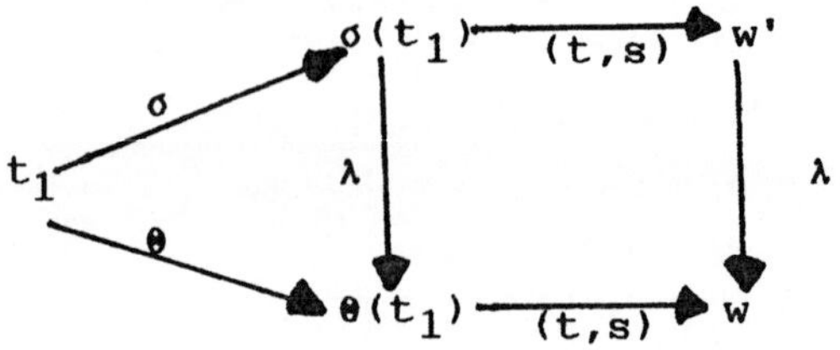

for some term w'. Therefore it is sufficient to consider (up to standardizing substitutions) most general substitutions σ of terms t and r where t is a left side of a reduction and r is a subterm of a left side of a reduction.
This means we have only to test the finitely many critical pairs of **R**.

[]

According to the last theorem we have a method to find t, t_1 and t_2 such that $t \;_{\mathbf{R}}\!\!\rightarrow^* t_1$, $t \;_{\mathbf{R}}\!\!\rightarrow^* t_2$, $t_1 \neq t_2$ and t_1, t_2 are both irreducible in case **R** does not have CR (but enjoys FTP and is finite). Of course, **R** implies the equation $t_1 \equiv t_2$ semantically. Therefore it is a natural idea to add one of the reductions $t_1 \rightarrow t_2$ or $t_2 \rightarrow t_1$ to the original system **R**. This should be done in such a way that the new system enjoys FTP again; then one wants to iterate the procedure.

Although there are only finitely many critical pairs to begin with this process might not stop because the added reductions give rise to new critical pairs. To make this more precise we assume a well-founded term ordering "$\prec$" and assume that **R** is finite and compatible with "$\prec$".
The following is called an algorithm although it is really a general frame for algorithms.

3. Definition:

The (redundant) Knuth-Bendix completion algorithm ($KBCA^r$) is defined by:

(i) $R_o^r := R$

(ii) Form all critical pairs (w_1, w_2) of R_n^r.

(iii) $P_n^r := \{t_1 \to t_2 \mid t_2 \prec t_1,\ t_1, t_2 \in Irred(w_j, R_n^r),$
$j=1,2,\ (w_1,w_2)$ a critical pair$\}$

(iv) If $P_n^r = \emptyset$, then the algorithm stops with result R_n^r, otherwise put $R_{n+1}^r := R_n^r \cup P_n^r$ and return to (ii).

Here we have neglected substitutions which separate variables. The algorithm generates many unnecessary reductions. We will give precise notions to deal with this.

4. Definition:

(i) $s \to t \in R$ is redundant in R iff

a) $s = t$ (s and t are identical)

or b) $t \in Red(R)$

or c) $s \in Red(R \setminus \{s \to t\})$.

(ii) R is redundant iff R contains a redundant reduction.

Redundant reductions cannot simply be omitted. They can, however, be replaced by "simpler" reductions. The following steps do change the set of reducible terms and leave the equational theory invariant:

1) If $s \to t \in R$ and $t \xrightarrow[R]{*} t'$, then replace $s \to t$ by $s \to t'$.
2) If $s \to t \in R$, $s \xrightarrow[R]{*} s'$, then replace $s \to t$ by $s' \to t$ or $t \to s'$ according whether $t \prec s'$ or $s' \prec t$.
3) Omit $s \to s$.

These steps can be incorporated in various ways into an algorithm which removes redundancies. In case **R** was already complete the procedure can be simplified:
Instead of 1) and 2) we take

1') If $s \to t \in \mathbf{R}$ and $t \;_{\mathbf{R}}\!\!\to t'$, then replace $s \to t$ by $s \to t^*$ where $t' \;_{\mathbf{R}}\!\!\to^* t^*$ and $t^* \in Irr(\mathbf{R})$, i.e. $t^* \in Irred(t',\mathbf{R})$.

2') If $s \to t \in \mathbf{R}$ and $s \;_{\mathbf{R}}\!\!\to s'$, then omit $s \to t$.

Again this can be incorporated into an algorithm. Observe, however, that the steps cannot be carried out parallel.

Starting with a complete reduction system $\mathbf{R}^r$ the non-redundant complete system **R** has the following properties:

If $s \to t \in \mathbf{R}$, then
a) $s \in Left(\mathbf{R}^r)$
b) $t \in Irr(\mathbf{R}^r)$
c) $s \;_{\mathbf{R}^r}\!\!\to^* t$
d) $s \in Irr(\mathbf{R}\backslash\{s \to t\})$.

Next we will modify the completion algorithm in order to generate only non-redundant reductions.

5. Definition:

The non-redundant KB-completion algorithm (KBCA) defines two sequences $\mathbf{R}_n, \mathbf{P}_n$ in the same way as $\mathbf{R}_n^r$, $\mathbf{P}_n^r$ except that (iv) is replaced by

(iv') If $\mathbf{P}_n = \emptyset$, stop with result $\mathbf{R}_n$, otherwise put
$$\mathbf{R}_{n+1} = \{s_1 \to s_2 \mid s_2 \prec s_1,\ s_j \in Irred(t_j, \mathbf{R}_n(t_1,t_2))\ j=1,2 \text{ where } t_1 \to t_2 \in \mathbf{R}_n \cup \mathbf{P}_n \text{ or } t_2 \to t_1 \in \mathbf{R}_n \cup \mathbf{P}_n \text{ and } \mathbf{R}_n(t_1,t_2) = \mathbf{R}_n\backslash\{t_1 \to t_2, t_2 \to t_1\}\ \}.$$

6. Definition:

The completion algorithm continues sucessfully if for each $n \in \mathbf{N}$ and each critical pair (w_1, w_2) all $t_1, t_2 \in \mathrm{Irred}(w_j, \mathbf{R}_n)$, $j=1,2$, $t_1 \neq t_2$, are comparable.

The completion algorithm may terminate for two reasons:

(a) For each critical pair (w_1, w_2) one has
$$\mathrm{Irred}(w_1, \mathbf{R}_n) = \mathrm{Irred}(w_2, \mathbf{R}_n).$$
This is called a successful termination.

(b) $\mathrm{Irred}(w_1, \mathbf{R}_n) \neq \mathrm{Irred}(w_2, \mathbf{R}_n)$ for some critical pair (w_1, w_2) but no terms in these sets are comparable with respect to "$\prec$". This is called a failure termination.

Successful termination results in complete sets while failure termination produces an incomplete set. If the completion algorithm continues succesfully then failure termination does not occur.

The main difference between the sequences $(\mathbf{R}^r_n, n \in \mathbf{N})$ and $(\mathbf{R}_n, n \in \mathbf{N})$ is that the first is increasing which the latter in general is not. For this reason we have to define the "limit systems" in a different way.

7. Definition:

(i) $\mathbf{R}^r := \cup (\mathbf{R}^r{}_n \mid n \in \mathbf{N})$

(ii) $\mathbf{R}^c := \cup (\mathbf{R}_n \mid n \in \mathbf{N})$

(iii) $\mathbf{R}^\infty := \{s_1 \to s_2 \mid s_j \in \mathrm{Irred}(t_j, \mathbf{R}^c(t_1, t_2)),\ j = 1,2,$ where $t_1 \to t_2 \in \mathbf{R}^c$ or $t_2 \to t_1 \in \mathbf{R}^c$ and $\mathbf{R}^c(t_1, t_2) = \mathbf{R}^c \setminus \{t_1 \to t_2, t_2 \to t_1\}\}$

$\mathbf{R}^\infty$ is called the limit system of the original system $\mathbf{R}$ which respect to "$\prec$".

<u>8. Proposition:</u>

Assume that the completion algorithm continues succesfully. Then $\mathbf{R}^r$, $\mathbf{R}^c$ and $\mathbf{R}^\infty$ are all complete systems and $\mathbf{R}^\infty$ is non-redundant.

<u>Proof:</u> By construction all three systems enjoy FTP. The completeness of $\mathbf{R}^r$ is immediate. By the remark above, replacing redundant reductions does not change the set of reducible terms, therefore the irreducible terms remain the same too. Hence induction gives $\mathrm{Irr}(\mathbf{R}^r_n) = \mathrm{Irr}(\mathbf{R}_n)$ for each $n \in \mathbf{N}$. This implies that $\mathbf{R}^c$ is complete. Because for each term t there are only finitely many t such that s is derivable from t. The irreducibles derivable from t in $\mathbf{R}^\infty$ can be obtained in some $\mathbf{R}_n$ already; this implies that $\mathbf{R}^\infty$ is complete. By definition $\mathbf{R}^\infty$ is non-redundant.

[]

The sets $\mathbf{R}_n$ are the <u>approximations</u> or <u>stages</u> of $\mathbf{R}^\infty$. As we will see, $\mathbf{R}^\infty$ can be infinite. This fact also follows from the undecidability of the word problem (e.g. for semigroups or groups). On the other hand, $\mathbf{R}^\infty$ depends only on the original system $\mathbf{R}$ and the ordering "$\prec$". We will discuss this and related questions in chapter II.4..

In the ground case the completion algorithm is particulary simple. Suppose we have a total term ordering "$\prec$" on the ground terms. Then $t \in \mathrm{superpos}(t_1,t_2)$ implies that t is a subterm of t_1. The critical pair diagram looks as follows:

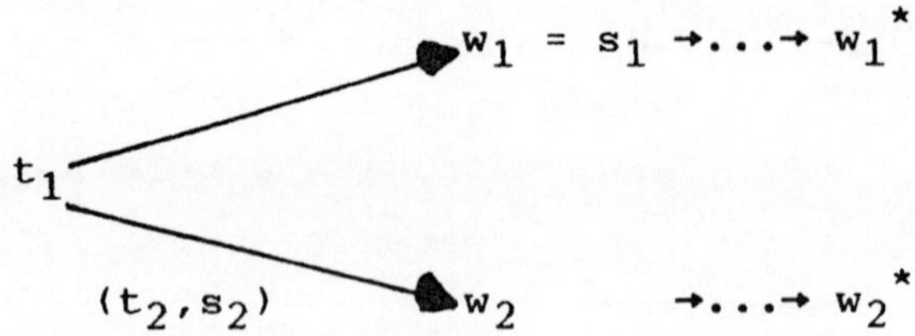

where w_1^* and w_2^* are irreducible. The only reductions which the completion algorithm possibly adds are $w_1^* \to w_2^*$ or $w_2^* \to w_1^*$; in both cases we have $w_1^* \prec t_1$ and $w_2^* \prec t_1$. Because "$\prec$" has no infinite descending chains there are only finitely many such reductions, i.e. the completion algorithm terminates (succesfully).

Later on we will consider a modified situation. There we consider group or semigroup words and will redefine the completion algorithm for such a situation. Although the completion algorithm deals only with ground words it will not terminate in general. This comes from the fact that general equations are implicitly present; e.g. the law of associativity is replaced by words written without parenthesises.

Finally we will give a (classical) example which arises from the axioms of group theory.
These are

$$E = \{e \cdot x = x,\ x^{-1} \cdot x = e,\ (x \cdot y) \cdot z = x \cdot (y \cdot z)\}.$$

In order to define a KB-ordering the following weights are choosen:

$$\omega_1 = \omega(e) = 1$$
$$\omega_2 = \omega(\cdot) = 0$$
$$\omega_3 = \omega(\cdot^{-1}) = 0$$

The initial stage for the completion algorithm is

$$\mathbf{R} = \mathbf{R}_o = \{e \cdot x \to x,\ x^{-1} \cdot x \to e,\ (x \cdot y) \cdot z \to x \cdot (y \cdot z)\}.$$

$\mathbf{R}_o$ does not have CR, as e.g. the following critical pair $(e \cdot y,\ (x^{-1} \cdot x) \cdot y)$ for $(x^{-1} \cdot x) \cdot y \in \text{superpos}((x \cdot y) \cdot z,\ x^{-1} \cdot x))$ shows:

We have:

$$(x^{-1} \cdot x) \cdot y \in superpos((x \cdot y) \cdot z, \ x^{-1} \cdot x))$$

$e \cdot y \qquad x^{-1}(x \cdot y)$

y

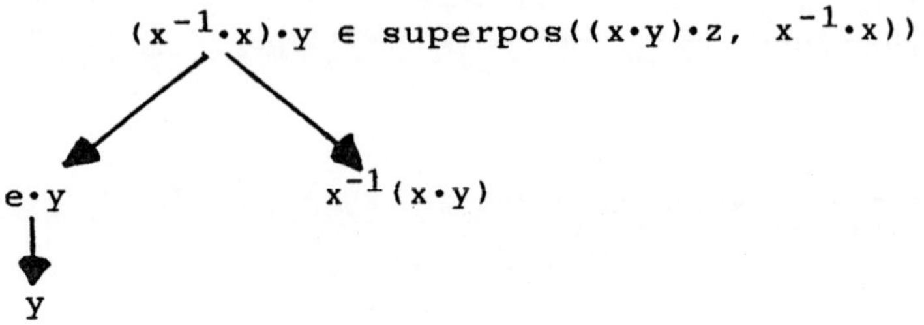

The completion algorithm generates as a complete system the reductions:

$$e \cdot x \to x \qquad\qquad e^{-1} \to e$$
$$x^{-1} \cdot x \to e \qquad\qquad (x^{-1})^{-1} \to x$$
$$(x \cdot y) \cdot z \to x \cdot (y \cdot z) \qquad x \cdot x^{-1} \to e$$
$$x^{-1}(x \cdot y) \to y \qquad x \cdot (x^{-1} \cdot y) \to y$$
$$x \cdot e \to x \qquad\qquad (x \cdot y)^{-1} \to y^{-1} \cdot x^{-1}$$

Additional remarks:

We do not intend to present an historical account of the concept of a complete system of reductions, the critical pair notion and the various forms of the completion algorithm. These ideas have been treated (to a large extent independently) by many different authors. An early appearance of the Church-Rosser property is in [New 42] which was applied to the word problem in [Ev 51]. A breakthrough for general algebraic structures was done 1970 in [Kn-Be] which also had much influence on the terminology. For the case of polynomial rings a complete system corresponds to a "Gröbner base"; completion was here first studied in Buchberger's thesis [Bu 65]. We also refer to [Ba-La 81] for commutative semigroups. More historical comments can be found in [La 81] and [Ba-Bu-La].

The notions of CR and WCR have a great number of variations which are useful for different purposes. The terminology in this area is not uniform. A detailed discussion of this and related topics is in [Hu 80] and [Hu-Op].

There are various methodes to ensure FTP which are all connected to certain partial or total well-founded orderings. We mention [Kn-Be], [La 75], [La 79], [Der-Ma], [Jou-Le-Re], [Ka-Na-Si]. A good overview is given in [Der 85].

Additional equations have also been "built in" in the completion procedures of e.g. [Pe-Sti].

II.3. The Ground Case

In the case of finitely generated groups one works in principle with two types of equations:

1) The general axioms of group theory;
2) The specific (ground) equations for the group under consideration.

Similar situations occur frequently as in semigroups, Boolean algebras etc.. The general equations simplify the situation considerably; they allow e.g. to leave the term algebra and to present terms by group words. It is of course desirable to avoid the explicit appearance of the general equations in the KB-completion algorithm too. This means that one has to "built-in" the effect of the general equations into some form of a special completion algorithm which works on the special equations only. This is in particular promising if one has a complete system for the general equations.

Let us first of all leave the term algebra by means of a special completion algorithm for finitely presented semigroups working on words, so that there is no longer need to care about the law of associativity.
As will be clear later, this semigroup completion algorithm is more efficient and easier to handle and implement although it is defined parallel to the KB completion algorithm.

Being precise we recall some formal definitions.

1. Definition:

Let $\Sigma = \{a_1,\ldots,a_n\}$ be a finite alphabet Σ^+ denotes the free semigroup over Σ and Σ^* the free monoid (i.e. the free semigroup with identity e) over Σ.
The elements $w \in \Sigma^+$, $w = w_1 w_2 \ldots w_k$ with $w_i \in \Sigma$, $1 \le i \le k$, are called words over Σ.
The identity e is also called the empty word.
Two words $u := u_1 \ldots u_k$ and $v := v_1 \ldots v_m$, $u_i, v_j \in \Sigma$ for $1 \le i \le k$, $1 \le j \le m$ are equal in Σ^*, denoted by $u = v$ iff $k=m$ and $u_i = v_i$ for all $i \in \{1,\ldots,k\}$.

2. Definition:

Let Σ be a finite alphabet.
A set $\mathbf{R} \subseteq \Sigma^* \times \Sigma^*$ of ordered pairs $\langle u,v \rangle$ of words is called a reduction system over Σ^* (often also called Thue-system over Σ^*); its elements, the reductions, will be denoted by $u \to v$.

Nearly all technical terms connected to reduction sytems over term algebras can be defined literally for reduction systems over Σ^*. For reasons of completeness they are listed here again.

3. Definition:

Let $\mathbf{R}$ be a reduction sytem over Σ^*.

(a) A word $w_1 \in \Sigma^*$ is an immediate reduction of a word $w_o \in \Sigma^*$ by $u \to v \in \mathbf{R}$ (denoted by $w_o \;{}_{u\to v}\!\!\to w_1$) iff there are $x,y \in \Sigma^*$ such that $w_o = xuy$ and $w_1 = xvy$.

(b) $w_o \;{}_{\mathbf{R}}\!\!\to w_1$ means $w_o \;{}_{u\to v}\!\!\to w_1$ for some $u \to v \in \mathbf{R}$.

(c) $R^{\rightarrow *}$ is the reflexive and transitive hull of $R^{\rightarrow}$.

(d) $w_o\ R^{\leftrightarrow}\ w_1$ means $w_o\ R^{\rightarrow}\ w_1$ or $w_1\ R^{\rightarrow}\ w_o$.

(e) $R^{\leftrightarrow *}$ is the reflexive and transitive hull of $R^{\leftrightarrow}$, i.e. $R^{\leftrightarrow *}$ is the <u>congruence</u> <u>relation</u> <u>generated</u> <u>by</u> **R**. Thus two words $v,w \in \Sigma^*$ are called <u>congruent</u> <u>modulo</u> **R** iff $v\ R^{\leftrightarrow *}\ w$.

Notation: $w \equiv v(\text{mod } \mathbf{R})$

The congruence class of w modulo **R** is again denoted by $[w]_{\mathbf{R}}$.

(f) For every $w \in \Sigma^*$ we put $L(w,\mathbf{R}) := \{v \mid w\ R^{\rightarrow *}\ v\}$.

(g) A word w is called <u>irreducible</u> with respect to **R** iff $L(w,\mathbf{R}) = \{w\}$.

(h) <u>Irr</u> = <u>Irr(R)</u>:= $\{w \in \Sigma^* \mid w$ is irreducible$\}$

<u>Red</u> = <u>Red(R)</u>:= $\{w \in \Sigma^* \mid w$ is reducible$\}$.

(i) <u>Irred(w,R)</u> = $L(w,\mathbf{R}) \cap Irr$

<u>Irred(S,R)</u> = $\bigcup_{w \in S} Irred(w,\mathbf{R})$ for $S \subseteq \Sigma^*$

(j) **R** has the <u>finite</u> <u>termination</u> <u>property</u> <u>FTP</u> iff there are no infinite chains $t_n\ R^{\rightarrow *}\ t_{n+1}$, $t_n \neq t_{n+1}$, $n \in \mathbb{N}$.

(k) **R** has the <u>Church-Rosser</u> <u>property</u> <u>CR</u> iff for all $w \in \Sigma^*$ and $w_1, w_2 \in L(w,\mathbf{R})$ we have $L(w_1,\mathbf{R}) \cap L(w_2,\mathbf{R}) \neq \emptyset$.

(l) **R** has the <u>unique</u> <u>termination</u> <u>property</u> <u>UTP</u> iff **R** has FTP and Irred(w,**R**) contains exactly one term for each $w \in \Sigma^*$.

(m) **R** is <u>complete</u> iff **R** has FTP and CR.

<u>4. Lemma:</u>

Let **R** be a reduction system over Σ^* and $\langle \mathbf{R} \rangle = R^{\leftrightarrow *}$.

(a) The quotient $\Sigma^+/_{\langle \mathbf{R} \rangle}$ is a semigroup under the operation $[u] \cdot [v] = [uv]$.

(b) The quotient $\Sigma^*/_{\langle \mathbf{R} \rangle}$ is a semigroup with identity $[e]$, also called a monoid.

5. Definition:

Let **R** be a reduction system over Σ^* and $\langle R\rangle = R^{\leftrightarrow^*}$.

We call $\Sigma^+/_{\langle R\rangle}$ resp. $\Sigma^*/_{\langle R\rangle}$ the <u>semigroup</u> resp. the <u>monoid presented by R</u>, **R** its <u>presentation</u>, the elements $u \to v \in R$ its <u>relators</u> and the elements $a_i \in \Sigma$ its <u>generators</u>.
If **R** is finite we say $\Sigma^+/_{\langle R\rangle}$ resp. $\Sigma^*/_{\langle R\rangle}$ is <u>finitely presented</u>.
Notation: $\Sigma^+/_{\langle R\rangle} =: SG_w(\Sigma,R)$
$\Sigma^*/_{\langle R\rangle} =: M_w(\Sigma,R)$
By $SG_w(\Sigma)$ resp. $M_w(\Sigma)$ we also denote the free semigroup resp. the free monoid generated by Σ.

6. Definition:

Let $\Sigma = \{a_1,\ldots,a_k\}$ be a finite alphabet and let $\omega\colon \Sigma \to \mathbb{N}\setminus\{0\}$ be a function which assigns a weight to every $a_i \in \Sigma$.
A <u>KB-ordering</u> $\ll$ over Σ^* is then defined as follows:
For $u = u_1u_2\ldots u_n$ and $v = v_1v_2\ldots v_m$ with $u_i, v_j \in \Sigma$, $1\le i\le n$, $1\le j\le m$ is $u \ll v$ iff

$$\text{(i)} \quad \sum_{i=1}^{n} \omega(u_i) < \sum_{j=1}^{m} \omega(v_j)$$

$$\text{or (ii)} \quad \sum_{i=1}^{n} \omega(u_i) = \sum_{j=1}^{m} \omega(v_j)$$

and (a) $\omega(u_1) < \omega(v_1)$
or (b) $\omega(u_1) = \omega(v_1)$ and $k<l$ if $u_1=a_k$ and $v_1=a_l$
or (c) $u_1 = v_1$ and $u_2\ldots u_n \ll v_2\ldots v_m$.

Furthermore we have $e \ll v$ for all $v \in \Sigma^+$.

The fact that the empty word e, the identity of the monoid Σ^*, has no weight, does not meet the definition of the KB-orderings over term algebras, since the identity of a term algebra is always presented by an O-ary function. For that and some other reason, discussed later (Lemma 12), we will define (Definition 13) a special KB-ordering for semigroups over term algebras which corresponds to $\ll$.

7. Lemma:

Every KB-ordering over Σ^* is well-founded and total.

8. Lemma:

Every semigroup $SG_w(\Sigma,\mathbf{R}')$ has a presentation $\mathbf{R}$ which has FTP.

Proof: Let $\ll$ be some KB-ordering over Σ^*. Since $\ll$ is total we have for every $u \to v \in \mathbf{R}'$ $u \gg v$ or $v \gg u$.

Hence $\mathbf{R}' = \mathbf{R}_1 \cup \mathbf{R}_2$, where

$$\mathbf{R}_1 := \{u \to v \mid u \to v \in \mathbf{R}',\ u \gg v\}$$
$$\mathbf{R}_2 := \{u \to v \mid u \to v \in \mathbf{R}',\ v \gg u\}.$$

And for $\mathbf{R} := \mathbf{R}_1 \cup \{v \to u \mid u \to v \in \mathbf{R}_2\}$ we have $u \gg v$ for every $u \to v \in \mathbf{R}$ which is a sufficient condition for the finite termination property of $\mathbf{R}$ because $\ll$ is well-founded.
Since $\mathbf{R}$ and $\mathbf{R}'$ induce of course the same congruence relation everything is proved.

[]

Let us now define a completion algorithm for finitely presented semigroups $SG_w(\Sigma,\mathbf{R})$.

9. Definition:

Given a finitely presented semigroup $SG_w(\Sigma,\mathbf{R})$ such that $u \gg v$ for all $u \to v \in \mathbf{R}$ and some KB-ordering $\ll$.

(i) The semigroup completion algorithm $SGCA^r$ is defined by:

(1) $\mathbf{R}_o^r := \mathbf{R}$

(2) Form all critical pairs (w_1,w_2) of $\mathbf{R}_n^r$, i.e. form

$$\mathbf{CP}_n := \{(w_1,w_2) \mid (\exists\ l_1u \to r_1,\ ul_2 \to r_2 \in \mathbf{R}_n^r \text{ s.t. } w_1 = r_1l_2,\ w_2 = l_1r_2) \text{ or } (\exists\ ul_2v \to r_1,\ l_2 \to r_2' \in \mathbf{R}_n^r \text{ s.t. } w_1 = r_1,\ w_2 = ur_2v)\}$$

(3) $\mathbf{P}_n := \{(v_1,v_2) \mid v_1 \gg v_2 \text{ and } v_i \in Irred(w_j,\ \mathbf{R}_n^r),\ i,j=1,2,\ (w_1,w_2) \in \mathbf{CP}_n\}$

(4) If $\mathbf{P}_n = \emptyset$, then the algorithm stops with result $\mathbf{R}_n^r$;
otherwise put $\mathbf{R}_{n+1}^r := \mathbf{R}_n^r \cup \mathbf{P}_n$ and return to (2).

(ii) By SGCA we denote the non-redundant $SGCA^r$.

The critical pairs (r_1l_2,l_1r_2) resp. (r_1,ur_2v) correspond to the set of pairs which occurs in Nivat's criteria [Ni 70] for the completeness of Thue-systems. Hence a proof that the SGCA generates a complete reduction system can be easily found by considering these criteria.
But this is not our intention. What we will rather do here is to show that the SGCA can be viewed as a special version of the KB-completion algorithm.

So let us return to the term algebras and remember that a finitely presented semigroup can also be understood as the subalgebra of ground terms of a term algebra which is factored by the fully invariant congruence generated by the law of associativity (ass) and the finitely many equations of the presentation **E**.

In order to be precise and for technical purposes we will introduce the following notations.

10. Definition:

Let $\Sigma = \{a_1,\ldots,a_n\}$ be a finite alphabet and e a letter not in Σ.
Further let $\mathbf{A} = \langle A,\Sigma,f\rangle$ and $\mathbf{A_e} = \langle A_e,\Sigma,e,f\rangle$ be algebras of signature $\langle 0,\ldots,0,2\rangle$.

(a) By $\underline{\mathbf{T}_{SG}(\Sigma)} := \mathbf{A}(Var)$ resp. $\underline{\mathbf{T}_{M}(\Sigma)} := \mathbf{A_e}(Var)$ we denote the term algebras of $\mathbf{A}$ resp. $\mathbf{A_e}$.

(b) For a finite set **E** of equations over $\mathbf{T}_{SG}(\Sigma)$ resp. $\mathbf{T}_{M}(\Sigma)$ we denote by
$\underline{SG_T(\Sigma,\mathbf{E})} = \{w \mid w \in \mathbf{T}_{SG}(\Sigma)/_{\langle\langle ass,\mathbf{E}\rangle\rangle},\ w \text{ ground term}\}$
resp.
$\underline{M_T(\Sigma,\mathbf{E})} = \{w \mid w \in \mathbf{T}_{M}(\Sigma)/_{\langle\langle ass,\, f(x,e)=f(e,x)=x\rangle\rangle},\ w \text{ ground term}\}$
the finitely presented semigroup resp. the finitely presented monoid.

(c) For a free semigroup resp. a free monoid we simply write $\underline{SG_T(\Sigma)}$ resp. $\underline{M_T(\Sigma)}$ instead of $SG_T(\Sigma,\emptyset)$ resp. $M_T(\Sigma,\emptyset)$.

A completion algorithm (KBCA) for $SG_T(\Sigma,E)$ will then, of course, start with

$$R_o(SG_T(\Sigma,E)) := R(E) \cup \{f(f(x,y),z) \to f(x,f(y,z))\}$$

$$(R(E) := \{t \to s \mid t \equiv s \in E \text{ or } s \equiv t \in E \text{ and } t \prec s\} \text{ for some KB-ordering } \prec)$$

and analogously for $M_T(\Sigma,E)$ with the reduction system

$$R_o(M_T(\Sigma,E)) := R(E) \cup \{f(f(x,y),z) \to f(x,f(y,z)),\ f(e,x) \to x,\ f(x,e) \to x\}$$

In order to carry terms over to words we define:

11. Definition:

Let T and T_e be the carriers of $T_{SG}(\Sigma)$ resp. $T_M(\Sigma)$.

(a) The transition $\underline{TW_e}$: $T_e \to (\Sigma \cup Var)^*$ is recursively defined as follows:

$$TW_e(e) := e \quad \text{(the empty word)}$$
$$TW_e(a_i) := a_i \quad \text{for } a_i \in \Sigma$$
$$TW_e(x) := x \quad \text{for } x \in Var$$
$$TW_e(f(t_1,t_2)) := TW_e(t_1) \cdot TW_e(t_2) \quad \text{for } t_1,t_2 \in T_e$$

(b) $\underline{TW}$ is the restriction of TW_e to T, i.e. $TW : T \to (\Sigma \cup Var)^+$

Obviously TW and TW_e are isomorphisms between $SG_T(\Sigma)$ and Σ^+ resp. between $M_T(\Sigma)$ and Σ^*.

As already mentioned there is a little discrepancy between the KB-orderings $\prec$ for term algebras and the KB-orderings $\ll$ over Σ^*.

12. Lemma:

There are KB-orderings $\ll$ over Σ^*, such that there are no KB-orderings $\prec$ over a term algebra **T** with the property $TW(\mathbf{T}) = (\Sigma \cup Var)^+$ which correspond to $\ll$ in the sense that $t_1 \prec t_2$ implies $TW(t_1) \ll TW(t_2)$ for all ground terms of **T**.

Proof: Consider for instance the following KB-ordering "$\ll$" over $\{a_1,a_2,a_3\}^*$ which is determined by the weights $\omega(a_1) = \omega(a_3) = 1$, $\omega(a_2) = 2$.
Of course we have $a_2 \gg a_1a_1$ and $a_3a_3 \gg a_2$.

Let $\prec_1$ be a KB-ordering over $\mathbf{T} = \mathbf{A}(Var)$, $\mathbf{A} = \langle A,a_1,a_2,a_3,f \rangle$ determined by the same weights as $\ll$ and a total ordering of $\{a_1,a_2,a_3,f\}$ such that $a_1<a_2<a_3$ and $f<a_2$.
Then we have $f(a_1,a_1) \prec_1 a_2$ and $f(a_3,a_3) \prec_1 a_2$ so that no $\prec_1$ corresponds to $\ll$.

On the other hand, if $\prec_2$ is a KB-ordering defined as $\prec_1$ except that we now demand $f>a_2$, then we have

$$f(a_1,a_1) \succ_2 a_2 \quad \text{and} \quad f(a_3,a_3) \succ_2 a_2.$$

Hence there is no KB-ordering $\prec$ which is equivalent to $\ll$.

[]

Small variations of the definition of KB-orderings over term algebras suffice to remove the above discussed discrepancy.

13. Definition:

Let $\prec$ be a KB-ordering over $T_{SG}(\Sigma)$ determined by a weight function ω with $\omega(f)=0$ and $\omega(a_i)>0$ for all $a_i \in \Sigma$.

We call $\prec_{SG}$ a <u>SG_T-ordering</u> over $T_{SG}(\Sigma)$ relative to the weight function ω iff

(a) $t_1 \prec_{SG} t_2$ if $t_1 \prec t_2$ and $t_i \in \Sigma$ implies $t_j \in \Sigma$ for $\{i,j\}=\{1,2\}$

(b) $t_1 \prec_{SG} t_2$ if $\omega(t_1)=\omega(t_2)$ and
(i) $t_1 \in \Sigma$, $t_2=f(s_1,s_2)$ and $t_1 \prec_{SG} s_1$
or (ii) $t_2 \in \Sigma$, $t_1=f(s_1,s_2)$ and $s_1 \prec_{SG} t_2$.

The second discrepancy occuring from the fact that the identity e of a monoid $M_T(\Sigma)$ is presented in Σ^* by the empty word, can, of course, be removed by demanding $\omega(e) = 0$. Note that this is not allowed for KB-orderings over term algebras. But since for every $t \in T_M(\Sigma)$ there is exactly one $t' \in [t]_M$ where $M = \langle ass,\ f(x,e) = f(e,x) = x \rangle$ such that t' is either e-free or identical e, we can define M_T-orderings $\prec_M$ over $T_M(\Sigma)$ as follows.

14. Definition:

Let $\prec_{SG}$ be a SG-ordering over $T_{SG}(\Sigma)$ determined by the weight function ω with $\omega(f) = 0$ and $\omega(a_i) > 0$ for all $a_i \in \Sigma$.
We call $\prec_M$ a <u>M_T-ordering over $T_M(\Sigma)$</u> relative to the weight function ω iff

(a) for t_1,t_2 e-free we have $t_1 \prec_M t_2$ iff $t_1 \prec_{SG} t_2$

(b) $e \prec_M t$ for $t \neq e$ and $t \notin Var$

(c) $t_1 \prec_M t_2$ if $t_1' \prec_M t_2'$ for $t_i' \in [t_i]_M$.

Since for every substitution σ we have $t_1 \prec_{SG} t_2$ implies $\sigma(t_1) \prec_{SG} \sigma(t_2)$ and $t_1 \prec_M t_2$ implies $\sigma(t_1) \prec_M \sigma(t_2)$ the orderings $\prec_{SG}$ and $\prec_M$ are, of course, term orderings.

Furthermore we have

15. Lemma:

For every KB-ordering $\ll$ over Σ^* resp. Σ^+ there is a M_T-ordering $\prec$ over $T_M(\Sigma)$ resp. a SG_T-ordering $\prec_{SG}$ over $T_{SG}(\Sigma)$ such that

$$t_1 \prec_M t_2 \quad \text{implies} \quad TW_e(t_1) \ll TW_e(t_2)$$

$$\text{resp.} \quad t_1 \prec_{SG} t_2 \quad \text{implies} \quad TW(t_1) \ll TW(t_2).$$

Proof: Let $\prec_M$, $\prec_{SG}$ and $\ll$ be determined by the same weight function ω with $\omega(f) = 0$ and $\omega(a_i) > 0$ for $a_i \in \Sigma$. A careful inspection of the definitions of $\prec_M$, $\prec_{SG}$ and $\ll$ yields then the lemma.

[]

We are now ready to prove:

16. Theorem:

Given a finitely represented semigroup $SG_w(\Sigma,R)$ such that $u \gg v$ for all $u \to v \in R$ and some KB-ordering $\ll$.
Let $E(R) := \{t \equiv s \mid t,s$ ground terms of $T_{SG}(\Sigma)$ and $TW(t) \to TW(s) \in R\}$
and let $\prec_{SG}$ be the SG-ordering corresponding to $\ll$.
Of course, $SG_T(\Sigma,E(R))$ is a term algebra representation of $SG_w(\Sigma,E(R))$, i.e. we have $TW(SG_T(\Sigma,E(R))) = SG_w(\Sigma,R)$ and $t_1 \prec_{SG} t_2$ implies $TW(t_1) \ll TW(t_2)$.

Furthermore let $\mathbf{R}_i^W$ be the reduction system generated by the SGCA after i steps starting with $\mathbf{R}_o^W := \mathbf{R}$ and $\mathbf{R}_i^T$ be the reduction system generated by the (redundant) KB-completion algorithm as defined by Definition 3 of chapter II.2. after i steps starting with
$\mathbf{R}_o^T := \{f(f(x,y),z) \to f(x,f(y,z)),\ t \to s \mid t \equiv s \in E(\mathbf{R})\}$.

Then we have for all $i \in \mathbb{N}$
$\mathbf{R}_i^W = \{TW(t) \to TW(s) \mid t \to s \in \mathbf{R}_{i+1}^T,\ t,s \text{ ground terms}\}$.

<u>Proof:</u> The proof is by induction on i.

<u>i = 0:</u> First of all let us compute $\mathbf{R}_1^T$.
Since $\text{superpos}(\text{Left}(\mathbf{R}_o^T\backslash\{ass\}),\text{Left}(\mathbf{R}_o^T)) = \emptyset$ critical pairs can arise only from

$$\text{superpos}(\text{Left}(ass),t) = \begin{cases} f_{ass}(t,x) & \text{if } t=f(t_1,t_2)\in\text{Left}(\mathbf{R}_1^T\backslash\{ass\}) \\ f(f(f(x,y),z),w) & \text{if } t \in \text{Left}(ass) \end{cases}$$

where $f_{ass}(t,x) := \text{Irred}(f(t,x),ass)$.

The KB-algorithm then generates the following sets:

$$\mathbf{CP}_o^T = \{(f_{ass}(t,x),f_{ass}(s,x)) \mid t \to s \in \mathbf{R}_o^T\backslash\{ass\}\} \cup \{(f(f(x,f(y,z)),w),f(f(x,y),f(z,w)))\}$$

$$\mathbf{P}_o^T = \{ f_{ass}(t,x) \to f_{ass}(s,x) \mid t \to s \in \mathbf{R}_o^T\backslash\{ass\}\}$$

$$\mathbf{R}_1^T = \mathbf{R}_o^T \cup \mathbf{P}_o^T$$

Hence we obtain:

$$\begin{aligned} &\{TW(t) \to TW(s) \mid t \to s \in \mathbf{R}_1^T,\ t,s \text{ ground terms}\} \\ = &\{TW(t) \to TW(s) \mid t \to s \in \mathbf{R}_o^T,\ t,s \text{ ground terms}\} \\ = &\{TW(t) \to TW(s) \mid t \equiv s \in E(\mathbf{R}),\ t \succ_{SG} s\} \quad = \mathbf{R}_o^W. \end{aligned}$$

<u>$i \Rightarrow i+1$:</u> Let $G_i := \{t \to s \mid t \to s \in R_j^T,\ t,s$ ground terms$\}$;
and $Gx_j := \{f_{ass}(t,x) \to f_{ass}(s,x) \mid t \to s \in G_j\}$

We have already seen that $R_1^T = \{ass\} \cup G_o \cup Gx_o$.
More generally the following assertion holds:

<u>Assertion:</u> $R_j^T = \{ass\} \cup G_j \cup Gx_j$

Proof by induction on j:

<u>$j = 1$:</u> trivial

<u>$j \Rightarrow j+1$:</u> Obviously we have:

(a) superpos(Left(ass),Left(R_j^T)) generates Gx_j; hence it h no effect on P_i.

(b) superpos(Left(G_j),Left(R_j^T)) = Ø

(c) superpos(Left(Gx_j),Left(G_j)) contributes, if at all, ground reductions to P_j.
Let GP_j be that contribution to P_j.

(d) superpos(Left(Gx_j),Left(Gx_j)) contributes $\{f_{ass}(t,x) \to f_{ass}(s,x) \mid t \to s \in GP_j\} =: GxP_j$ to P_j.

(e) superpos(Left(Gx_j),Left(ass)) has no effect on P_j.

Hence we obtain

$$R_{j+1}^T = R_j^T \cup GP_j \cup GxP_j = \{ass\} \cup G_{j+1} \cup Gx_{j+1}$$

Suppose now $v \to w \in R_{i+1}^W \setminus R_i^W$.
Then there are $l_1 u \to r_1,\ u l_2 \to r_2 \in R_i^W$ such that

(a) $v \in Irred(r_1 l_2, R_i^W)$ and $w \in Irred(l_1 r_2, R_i^W)$

or (b) $v \in Irred(l_1 r_2, R_i^W)$ and $w \in Irred(r_1 l_2, R_i^W)$.

W.l.o.g. we assume case (a).
By the induction hypothesis there are

$f_{ass}(l_1^\sim, u^\sim) \to r_1^\sim$ and $f_{ass}(u^\sim, l_2^\sim) \to r_2^\sim \in G_{i+1}$

where $t^\sim \in TW^{-1}(t)$.

According to the preceding proposition there is also
$f_{ass}(l_1^\sim, f_{ass}(u^\sim,x)) \to f_{ass}(r_1^\sim,x) \in R_{i+1}^T$.

Hence $(f_{ass}(r_1^{\sim},l_2^{\sim}),f_{ass}(l_1^{\sim},r_2^{\sim}))$ is the critical pair belonging to

$$f_{ass}(l_1^{\sim},f_{ass}(u^{\sim},l_2^{\sim})) \in superpos(f_{ass}(l_1^{\sim},f_{ass}(u^{\sim},x)),\; f_{ass}(u^{\sim},l_2^{\sim}))$$

By a second application of the induction hypothesis we can conclude then, that there are $v^{\sim},w^{\sim}$ such that

$$v^{\sim} \in Irred(f_{ass}(r_1^{\sim},l_2^{\sim}),\mathbf{R}^T_{i+1}) \quad \text{and}$$
$$w^{\sim} \in Irred(f_{ass}(l_1^{\sim},r_2^{\sim}),\mathbf{R}^T_{i+1})$$

Hence $v^{\sim} \in w^{\sim} \in \mathbf{G}_{i+2} \subseteq \mathbf{R}^T_{i+2}$ what was to be proved.

Now let $t \to s \in \mathbf{G}_{i+2}\backslash\mathbf{R}^T_{i+1}$.
Then there is (see the proof of the preceding proposition) a $p \in superpos(Left(\mathbf{G}x_{i+1}),Left(\mathbf{G}_{i+1}))$ such that

(a) $t \in Irred(p_1,\mathbf{R}^T_{i+1})$ and $s \in Irred(p_2,\mathbf{R}^T_{i+1})$ or
(b) $t \in Irred(p_2,\mathbf{R}^T_{i+1})$ and $s \in Irred(p_1.\mathbf{R}^T_{i+1})$

where (p_1,p_2) is the critical pair belonging to p.
Again we stick only to case (a); case (b) can be proved analogously.
To be more precise there are however

$$f_{ass}(l_1,f_{ass}(u,x)) \to f_{ass}(r_1,x), \quad f_{ass}(u,l_2) \to r_2 \in \mathbf{R}^T_{i+1}$$

such that

$$p=f_{ass}(l_1,f_{ass}(u,l_2)), \quad p_1=f_{ass}(r_1,l_2), \quad p_2=f_{ass}(l_1,r_2).$$

Since there is also $f_{ass}(l_1,u) \to r_1 \in \mathbf{R}^T_{i+1}$ we can conclude by the induction hypothesis that
$TW(l_1)TW(u) \to TW(r_1)$ and $TW(u)TW(l_2) \to TW(r_2) \in \mathbf{R}^W_i$.

Hence $(TW(r_1)TW(l_2),TW(l_1)TW(r_2))$ is the critical pair belonging to $TW(p) = TW(l_1)TW(u)TW(l_2)$.

A second application of the induction hypothesis yields then of course
$TW(t) \in Irred(TW(p_1),R_i^W)$ and $TW(s) \in Irred(TW(p_2),R_i^W)$.

Hence $TW(t) \to TW(s)$ is generated by the $SGCA^r$ during the i-th step i.e. $TW(t) \to TW(s) \in R_{i+1}^W$.

[]

As we have seen above, there is a completion algorithm for semigroups which does no longer work explicitly with the law of associativity, by means of

(a) omitting brackets, i.e. turning to words instead of terms

and

(b) changing the CR-test so that the process of building the superpos-sets is replaced by the more simpler process of overlapping, which can be done by string matchings.

In the following we will do even more and define a completion algorithm for finitely presented groups with 'built-in' group axioms.
Again instead of terms over the free group generated by Σ we will treat words over $(\Sigma \cup \Sigma^{-1})^*$, where $\Sigma^{-1}=\{a^{-1} | a\in\Sigma\}$.

Analogously to the preceding terminology we fix our notations:

17. Definition:

Let $\mathbf{R}$ be a reduction system over $(\Sigma\cup\Sigma^{-1})^*$,
$FR := \{aa^{-1} \to e,\ a^{-1}a \to e \mid a\in\Sigma\}$ and $\langle \mathbf{R}\cup FR\rangle = {}_{\mathbf{R}\cup FR}{\leftrightarrow}^*$

We call $G_W(\Sigma,\mathbf{R}) := (\Sigma\cup\Sigma^{-1})^*/_{\langle \mathbf{R}\cup FR\rangle}$ the group presented by $\mathbf{R}$, $\mathbf{R}$ its representation, the elements $u \to v \in \mathbf{R}$ its relators and the element $a \in \Sigma$ its generators.
The free group generated by Σ is of course $(\Sigma\cup\Sigma^{-1})^*/_{\langle FR\rangle}$ and will be denoted by $\underline{G_W(\Sigma)}$.

Again we assume throughout the rest of this paragraph that the presentation $\mathbf{R}$ has FTP by being compatible to some well-founded ordering over $G_W(\Sigma)$. In particular, we suppose compatibility to a KB-ordering $\ll$ over $(\Sigma\cup\Sigma^{-1})^*$, determined by a weight function ω: $\Sigma\cup\Sigma^{-1} \to \mathbb{N}\setminus\{0\}$ with the property $\omega(a_i)=\omega(a_i^{-1})$ for all $a_i\in\Sigma$.

In accordance to Definition 17 the SGCA starting with $\mathbf{R}_0 = \mathbf{R}\cup FR$ is a handy completion algorithm for finitely presented groups $G_W(\Sigma,\mathbf{R})$. Thus again we can leave the term algebra and work on group words only.

In the group case one can achieve an even more efficient completion algorithm which takes advantage of the group axioms. As will be shown in the following, great parts of the SGCA for groups can be replaced by a very simple procedure called "symmetrizing".

To be precise we must establish some notations.

<u>18. Definition:</u>

Let **Q** be a subset of **R**.

(i) By <u>$PA^r(Q,R)$</u> we denote the following part of the completion algorithm for semigroups $SGCA^r$:

(1) $R_o^r := R \setminus Q$

(2) Form the following critical pairs by overlapping the left side of a reduction from R_n with some left side of a reduction of **Q**, i.e. form

$$CP_n := \{(w_1,w_2) \mid \exists\ l_1u \to r_1 \in R_n^r,\ ul_2 \to r_2 \in Q \text{ or } \exists\ l_1u \to r_1 \in Q,\ ul_2 \to r_2 \in R_n^r \text{ s.t. } w_1 = r_1l_2 \text{ and } w_2 = l_1r_2 \text{ or } \exists\ ul_2v \to r_1 \in R_n^r,\ l_2 \to r_2 \in Q \text{ or } \exists\ ul_2v \to r_1 \in Q,\ l_2 \to r_2 \in R_n^r \text{ s.t. } w_1 = r_1 \text{ and } w_2 = ur_2v\}$$

(3) $P_n := \{(v_1,v_2) \mid v_1 > v_2,\ v_i \in Irred(w_j, R_n^r \cup Q)\ i,j=1,2,\ (w_1,w_2) \in CP_n\}$

(4) If $P_n = 0$, then the algorithm stops with result R_n^r, otherwise put $R_{n+1}^r := R_n^r \cup P_n$ and return to (2).

(ii) By <u>$PA(Q,R)$</u> we denote the non-redundant $PA^r(Q,R)$.

Notice, in particular, that $PA^r(FR, R \cup FR)$ is the partial $SGCA^r$ which tests the CR-property only for critical pairs which arise by overlapping a left side of a reduction not in FR with some left side of a reduction of FR.

As it will be postulated in Theorem 22 the set of reductions generated by the PA(FR,**R**∪FR) is just Sym(**R**), defined below, if no "undesirable" reductions are possible.

19. Definition:

For $l \to r \in \mathbf{R}$ we define

(i) Sym($l \to r$) = $\{\lambda \to \varrho \mid \lambda \gg \varrho$, $\lambda\varrho^{-1}$ or $\lambda^{-1}\varrho$ is a cyclic permutation of lr^{-1} and for $\lambda = a\lambda'b$ we have $a\lambda' \ll \varrho b^{-1}$ and $\lambda'b \ll a^{-1}\}$

(ii) Sym(**R**) = $\cup$ (Sym($l \to r$)| $l \to r \in \mathbf{R}$).

In order to avoid "undesirable" reductions during the PA(FR,**R**∪FR) the following so called small cancellation conditions are very useful.

20. Definition:

Let **S** be a set of group words, i.e. $\mathbf{S} \subset (\Sigma\cup\Sigma^{-1})^*$ and λ a positive real number.

S satisfies the small cancellation condition C'(λ) iff wu, $wv \in \mathbf{S}$ implies $|w| \le \lambda|wv|$ for the lenght function $|..|$.

21. Definition:

(i) A group word $w = b_1 \ldots b_m$, $b_i \in \Sigma \cup \Sigma^{-1}$ is called cyclically reduced, if w is irreducible with respect to FR and b_m is not the inverse of b_1.

(ii) A set of group words **S** is symmetrized if all elements of **S** are cyclically reduced and all cyclic permutations of w or w^{-1} are in **S** for any $w \in$ **S**. (Let $w = b_1 \ldots b_m$ then of course $w^{-1} = c_m \ldots c_1$, with $c_i = a_j$ if $b_i = a_j^{-1}$ and $c_i = a_j^{-1}$ if $b_i = a_j$, where $a_i \in \Sigma$.)

(iii) The symmetrized closure of a set of group words **S**, i.e. the smallest set that is symmetrized and includes **S**, is denoted by Sym(**S**).

22. Theorem:

Let **S** := { lr^{-1} | $l \to r \in$ **R** } for a finitely presented group $G_w(\Sigma,$**R**$)$.
If Sym(**S**) satisfies the C'(1/2)-condition and $\ll$ is defined by $\omega(a) = \omega(a^{-1}) = 1$ for all $a \in \Sigma$, then Sym(**R**) is the result of PA(FR,**R**$\cup$FR).

The proof can be found analoguously to the proof of Theorem 16 by carrying out the PA(FR,**R**$\cup$FR).

Theorem 22 yields now the following completion algorithm for groups:

23. Definition:

Given a finitely represented group $G_W(\Sigma,\mathbf{R})$ such that $l \gg r$ for all $l \to r \in \mathbf{R}$ and some KB-ordering $\ll$.

(i) The group completion algorithm GCA^r is defined by:

(1) $\mathbf{R}_o^r := \mathrm{Sym}(\mathbf{R})$

(2) Form all critical pairs (w_1,w_2) of $\mathbf{R}_n^r$, i.e. form

$$\mathbf{CP}_n := \{(w_1,w_2) \mid \exists\ l_1u \to r_1,\ ul_2 \to r_2 \in \mathbf{R}_n^r, \text{ s.t. } w_1=r_1l_2 \text{ and } w_2=l_1r_2 \text{ or } \exists\ ul_2v \to r_1,\ l_2 \to r_2 \in \mathbf{R}_n^r, \text{ s.t. } w_1=r_1 \text{ and } w_2=ur_2v\ \}$$

(3) $$P_n := \{(v_1,v_2)/\ v_1 \ll v_2 \text{ and } v_i \in \mathrm{Irred}(w_j,\mathbf{R}_n^r),\ i,j=1,2,\ (w_1,w_2) \in \mathbf{CP}_n\ \}$$

(4) If $P_n=0$, then the algorithm stops with result $\mathbf{R}_n^r$, otherwise put $\mathbf{R}_{n+1} := \mathbf{R}_n^r \cup \mathrm{Sym}(P_n)$

(ii) Again we call GCA the non-redundant GCA^r.

Additional remarks:

As mentioned at the end of chapter I.1., there are various methods to incorporate additional axioms into the unification procedure. There are also a number of different methods to use additional laws in the ground case. As important examples we mention [Ni 70] for monoids, [Ba-La 81] for abelian semigroups and the work of Buchberger on (polynomial) rings, see e.g. [Bu 84]. Of particular interest is the case of Boolean algebras because of its relevance to logic. This seems first to be handled in [Hs 82],cf.also [Ka-Na85].

II.4. First Analysis of the Completion Algorithm

In chapter II.2. we defined the completion algorithm and its stages or approximations. In this section we restrict ourselves to the case where the completion algorithm continues successfully. Special attention is paid to the group and semigroup case. Throughout this section we assume a well-founded term ordering which is total on the ground terms. Successful continuation splits into two cases:

(I) Termination with a finite complete system $\mathbf{R}_n$.
(II) The limit system $\mathbf{R}^\infty$ is infinite.

$\mathbf{R}^\infty$ is also complete and reduces each term t to the $\prec$-minimal term t^* which is semantically equivalent to t. We want obtain different constructions of $\mathbf{R}^\infty$ and need to compare complete systems. The next proposition refers to notions in chapter II.1..

1. Proposition:

Two complete equationally equivalent systems **R** and **S** with the same irreducibles are derivation equivalent.

Proof: W.l.o.g. it suffices to show that $s \to t \in \mathbf{R}$ implies $s \; {}_{\mathbf{S}}\!\!\overset{*}{\to} t$. We can also assume that **R** is non-redundant. Then $t \in \mathrm{Irr}(\mathbf{R})$, hence $t \in \mathrm{Irr}(\mathbf{S})$. We also have some r such that

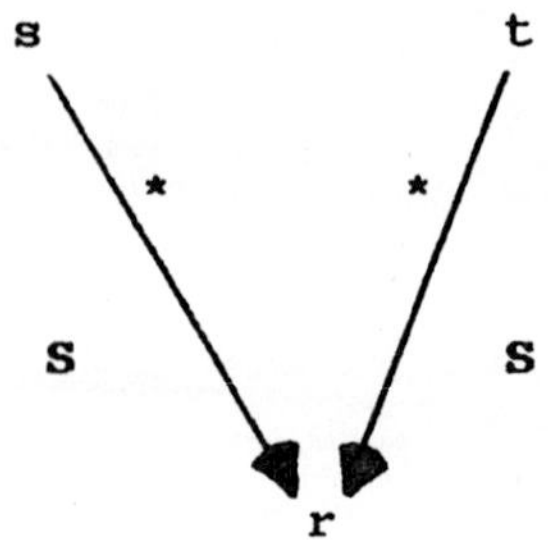

holds; $t \in Irr(S)$ then gives $t = r$ and we get $s \ {}_{S}{\to}^{*} \ t$.

[]

2. Proposition:

Suppose **R** and **S** are

(i) complete

(ii) non-redundant

(iii) either a) derivation equivalent

or b) equationally equivalent with the same irreducibles.

Then **R** = **S** up to a renaming of variables.

Proof: By the previous proposition and the fact that derivation equivalent systems have the same irreducibles we can assume that **R** and **S** are derivation equivalent and

$$Irr(\mathbf{R}) = Irr(\mathbf{S}) \quad \text{and} \quad Red(\mathbf{R}) = Red(\mathbf{S}) \quad \text{holds.}$$

We will show $\mathbf{R} \subseteq \mathbf{S}$ up to some renaming of variables.
Assume $s \to t \in \mathbf{R}$; then $s \in Red(\mathbf{R})$ and by non-redundancy $s' \notin Red(\mathbf{R})$ for all proper subterms s' of s.
By assumption we also have

$$s \ {}_{S}{\to} \ t_1 \ {}_{S}{\to} \ \cdots \ {}_{S}{\to} \ t_n = t$$

and obtain $s \ {}_{R}{\to}^{*} \ t_1$ which gives $t = t_1$ because **R** is non-redundant.

Therefore we have $u \rightarrow v \in \mathbf{S}$ and a substitution σ such that

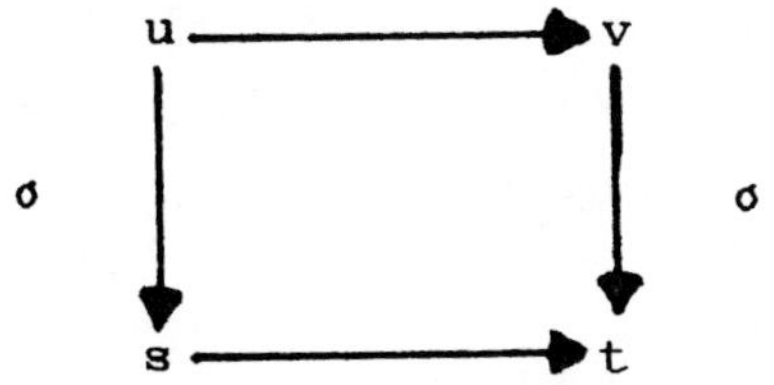

The same argument provides $s_o \rightarrow t_o \in \mathbf{R}$ and a substitution τ for which

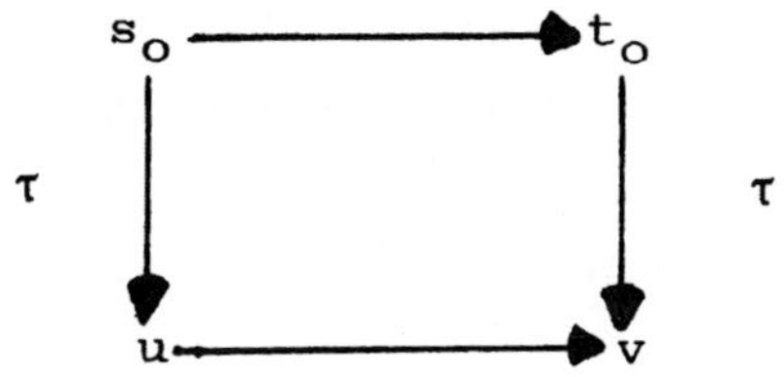

holds. This implies $s_o = s$, $t_o = s$; therefore σ and τ are renaming substitutions.

[]

3. Corollary:

Under the assumptions of the previous proposition **R** and **S** have the same ground reductions.

The argument also works for the group and semigroup case where we deal with words instead of terms.

A typical application is the construction of the limit system from "below" and from "above". Suppose **R** is a finite system of reductions and $\mathbf{R}^{\infty}$ is the limit system.

We also have

$$E = E(\mathbf{R}) = \{s \equiv t \mid s \rightarrow t \in \mathbf{R}\} \quad \text{and}$$
$$\mathbf{R'} = \{s \rightarrow t \mid t \prec s,\ E \Vdash s \equiv t\}$$
$$\text{(which is } \{s \rightarrow t \mid t \prec s,\ s \ {}_{\mathbf{R}}\!\leftrightarrow^{*} t\} \text{ in the semigroup case).}$$

By removing redundancies from $\mathbf{R'}$ we get a second non-redundant complete system $\mathbf{R}^{\sim}$ which is derivation equivalent to $\mathbf{R}^{\infty}$; hence we have obtained $\mathbf{R}^{\infty}$ from "above".
In particular, the process of removing redundancies results in a uniquely determined system. For the group and semigroup case where we deal with (ground) words only we state that for a given presentation the non-redundant complete system is uniquely determined by the chosen ordering; different orderings may however result in essentially different limit systems.
Running through all Knuth-Bendix orderings there are three possibilities for a given presentation **R**:

1) All limit systems are finite.
2) There are finite and infinite limit systems.
3) All limit systems are infinite.

As we will see, all possibilities can occur. First we consider finite algebras and fix the ordering $\prec$. For a finite algebra **A** with carrier A we put

$T(\mathbf{A}) := \{t \mid t \text{ a } \prec\text{-minimal term denoting some element in } A\}$.

Clearly $T(\mathbf{A})$ and **A** are in one-one correspondence.

For each n-ary operation f the <u>multipication</u> <u>table</u> is defined as the set of equations

$$M_f = \{f(t_1,..,t_n) \equiv t \mid t_1,..,t_n, t \in T(\mathbf{A}),\ \mathbf{A} \Vdash f(t_1,...,t_n) \equiv t\};$$

We put

$$M = \bigcup (M_f \mid f \text{ an operation}).$$

The multiplication table of a constant consists of just one equation.

If $\mathbf{R}_M = \{s \to t \mid s \equiv t \in M,\ s \prec t\}$

then all right sides of equations are irreducible; left sides are irreducible iff they coincide with the right sides. $\mathbf{R}_M$ is compatible with the ordering and therefore has FTP. Because each term t reduces to the minimal term t' for which $\mathbf{A} \Vdash t \equiv t'$ we have also UTP. Therefore $\mathbf{R}_M$ is complete.

Removing the redundancies in $\mathbf{R}_M$ is particularly simple. From the steps (1'), (2') and (3) in chapter II.2. step (1') does not apply because right sides are irreducible. Therefore the non-redundant complete system belonging to $\mathbf{R}_M$ is already a subsystem of $\mathbf{R}_M$.

Next we isolate a case where the completion algorithm terminates.

<u>4. Proposition:</u>

Suppose G is a finitely presented finite group or semigroup with an initial system $\mathbf{R}_o$. Then the completion algorithm terminates (i.e. $\mathbf{R}^\infty$ is finite) for all KB-orderings and there are only finitely many such systems.

<u>Proof:</u> Consider the complete limit system $\mathbf{R}^\infty$.
If $|G| = n$ then the maximum length of an irreducible word is at most n-1 because the irreducibles are closed under taking

subwords. Therefore the entries in the multiplication tables are words w with length $< n$ which means that there are only finitely many different tables. Hence there is a uniform bound $|\mathbf{R}_M| \leq k$ for all KB-orderings.
Therefore there is a uniform bound on $|\mathbf{R}^\infty|$ too, hence each $\mathbf{R}^\infty(\prec)$ is finite and there are only finitely many such non-redundant complete systems.

[]

These considerations are sometimes useful even when general equations with variables occur.

An example is given by the idempotent semigroups; the general equation describing idempotency is $x^2 = x$.
A key fact here is that each finitely generated idempotent semigroup is finite, in particular the free ones (see for example [Gr-Re]).
This is not quite obvious because for $n > 2$ there are infinitely many words on n letters which do not contain a square, i.e. a subword of the form uu.

An immediate consequence is

5. Proposition:

Each finitely generated idempotent semigroup has a finite complete system.

A way to construct the system is as follows:
We start with the reductions $AA \rightarrow A$ for each generator A. This gives a system which is complete but does not yet describe the semigroup.

We have two kinds of adding reductions:

1) Ordinary steps of the SGCA
2) Adding reductions of the form $uu \to u$ where u runs through all words of length $l \leq k$, where k is the maximal length of words $v \in \mathrm{Left}(\mathbf{R})$, $\mathbf{R}$ being the present system.

The termination conditions are

a) Completeness
b) Each word of length n, n the maximal length of words $v \in \mathrm{Left}(\mathbf{R})$, is reducible.

By the finiteness of the semigroup the algorithm stops.
In particular this gives a solution to the word problem in finitely generated idempotent semigroups; this question was discussed at several places in the literature (cf. [Gr-Re] or [Ger]).
A disadvantage of this method is its complexity. In particular the complete system is quite big, due to the fact that the size of the free idempotent semigroup increases sharply with the number of generators. The semigroup with 4 generators has e.g. already 332 388 many elements.
A more efficient procedure is the use of some conditional reduction system; this was carried out in [Sie-Sz]. We will discuss this below.

The next example will show that different orderings may in fact lead to different limit systems for finite algebras.
The Fibonacci-groups $F(2,n)$ on n generators $A_1,\ldots,A_n$ are defined by the equations

$$A_iA_{i+1} = A_{i+2} \quad \text{for } 1 \leq i \leq n-2,$$
$$A_{n-1}A_n = A_1$$
$$A_nA_1 = A_2$$

If we put $A_1 = A$, $A_2 = B$ and remove the remaining constants the equations for F(2,5) reduce to

$$ABBABA = B$$
$$BABABBA = A$$

For the first ordering we choose the weights

$$\omega(A) = 5$$
$$\omega(B) = 1$$

The complete system is:

$$A \rightarrow B^3$$
$$A^{-1} \rightarrow B^{-3}$$
$$B^{-6} \rightarrow B^5$$
$$B^6 \rightarrow B^{-5}$$

This system gives the immediate information that the F(2,5) is just $\mathbb{Z}_{11}$.

For the second ordering we take $\omega(A) = \omega(B) = 1$.

The complete system is:

$B^{-2} \rightarrow BA^{-1}$	$B^{-1}A^{-1} \rightarrow A^{-1}B^{-1}$
$B^2 \rightarrow AB^{-1}$	$A^{-2}B^{-1} \rightarrow AB$
$ABA^{-1} \rightarrow B$	$A^2B \rightarrow A^{-1}B^{-1}$
$A^{-1}B \rightarrow BA^{-1}$	$A^3 \rightarrow BA^{-1}$
$BA^{-1}B \rightarrow A^{-1}$	$BA^{-2} \rightarrow A^2$
$BA \rightarrow AB$	$A^2B^{-1} \rightarrow A^{-2}$
$B^{-1}A \rightarrow AB^{-1}$	$A^{-3} \rightarrow AB^{-1}$

This system is much less useful for an immediate overview. But note that the commutativity is expressed directly.

We now turn to situations where $\mathbf{R}^\infty$ can be infinite as well as finite, depending on the ordering.
There are various ways to describe infinite limit systems which we will investigate later. In order to prove that the limit system is infinite it often suffices to know sufficiently many reductions or irreducible elements.
One of the classical examples where the completion algorithm terminates was given in chapter II.2. where we presented the complete system for pure group theory. Already in [Kn-Be] it was shown that changing the ordering can result in a loss of the termination property for the completion algorithm. The new ordering was obtained from the old one by changing the weight for the inverse:

$$\omega((\ )^{-1}) = 1.$$

As a consequence of that several reductions have to be reversed, we obtain e.g.

$$y^{-1}x^{-1} \rightarrow (xy)^{-1}.$$

This suggests to consider the terms

$$t_n := (x_1(x_2(\ldots(x_{n-1}x_n)\ldots)))^{-1}\ , \quad n\geq 2$$

which are minimal words in their congruence classes (any other word s equal to t_n would contain all x_k but more occurrences of $(\ )^{-1}$).
If we take

$$s_n := x_{n+1}(x_1(x_2(\ldots(x_n x_{n+1})\ldots)))^{-1}, \quad n\geq 2$$

then for all n $s_n \equiv t_n$ holds and all proper subterms of s_n are minimal too. This means that we have for any complete system $\mathbf{R}$ compatible with this ordering:

$$s_n \ \xrightarrow[\mathbf{R}]{*}\ t_n$$

with $s \in Irr(R)$ for all proper subterms s of s_n.

Therefore for some $s \in Left(R)$ and some substitution σ we have $s_n = \sigma(s)$; σ must be a renaming of variables because otherwise the irreducible t_n would be a substitution instance of s too.
This implies that the limit system R^∞ must be infinite.

Another example are the idempotent semigroups; note that we deal with words now instead of terms. In [Sie-Sz] the following conditional and complete system was given for this theory:

(i) $xx \to x$

(ii) $xyz \to xz$ provided that $\{x\} = \{z\}$ and $\{xy\} = \{z\}$

(Here $\{u\}$ denotes the content of u, i.e. the set of variables and constants occurring in u.)

We view this conditional system as an ordinary infinite system (which is of course redundant). The non-redundant system is difficult to exhibit; in order to show that it is still infinite it again suffices to consider a certain infinite set of reductions.
We put

$$s_n := x_1x_2 \cdots x_nx_1x_n \cdots x_2x_1$$
$$r_n := x_1x_2 \cdots x_nx_n \cdots x_2x_1$$
$$t_n := x_1 \cdots x_{n-1}x_nx_{n-1} \cdots x_1$$

and get

$$s_n \to r_n \to t_n.$$

Furthermore we have for all $n \geq 2$:

(a) Each proper subword of s_n is irreducible

(b) s_n is reducible only by renamings of $s_n \to r_n$

(c) The only subword reducible in r_n is x_nx_n

(d) t_n is irreducible.

The technique of proof is in all cases the same and we restrict ourselves to a typical one. First one sees that all substitutions involved have to be renamings of variables. Suppose e.g. that a subword of s_n is reduced by a reduction of the form $xyz \to xz$ and x starts with some x_i, $i > 1$, in the first half of s_n. The z has to contain x_i and hence x_1 is in xyz. Therefore x_1 has to be in x, xy and z; because there are only two remaining occurrences of x_1 we see that x_1 is not in y. This means that y contains some other x_k which shows $\{x\} \neq \{z\}$; hence we have a contradiction.
It follows from (a) and (d) that the special reductions $s_n \to t_n$ cannot be omitted.
Note that this argument works for all KB-orderings because the weight of the variables is always the minimum weight of the constants. Combining this with our considerations from above we get for all KB-orderings

(1) The general theory of idempotent semigroups does not have a finite complete system.

(2) Each specific finitely generated idempotent semigroup admits a finite complete system.

Now consider the general case of an arbitrary ordering. Any reduction has to be length reducing because otherwise the special substitution which replaces x for every variable x_i would give a reduction $x^n \to x^m$ with $n \le m$ and violate FTP. Therefore any complete system is also compatible with the KB-orderings and hence infinite by the arguments from above.

We now turn to the ground case.
Suppose $G = \langle a_1,\dots,a_n \mid u \equiv e\rangle$ where some $a = a_i$ occurs exactly once, i.e. $u = vaw$ and a is not in vw. We choose weights such that $\omega(a) > \omega(v) + \omega(w)$. Then the completion algorithm generates

$$R^\infty = \{a \to v^{-1}w^{-1},\ a^{-1} \to wv\}.$$

We observe here that there is a connection to the "Freiheitssatz" of Magnus (cf. e.g. [Ly-Sch]). This says that for one relator groups each generator occurring in the defining relation occurs also in each non-trivial reduced word denoting the identity element.
Notice that a less skillful choice of weights may lead to an infinite limit sytem R^∞.

Finally we will give an example where the limit system R^∞ is inifinte for all KB-orderings.
Consider the so called Greendlinger group

$$G = \langle\ A,\ B,\ C\ |\ \ ABC \equiv CBA\ \rangle$$

For each weight function $\omega_{ABC} \in \{\ \omega\ |\omega$ is a weight function s.t. $\omega(A)<\omega(B)<\omega(C)\}$
the completion algorithm GCA generates:

$$\begin{aligned} R^\infty = \{ & CBA \to ABC,\ A^{-1}CB \to BCA^{-1},\ B^{-1}A^{-1}C \to CA^{-1}B^{-1} \\ & C^{-1}B^{-1}A^{-1} \to A^{-1}B^{-1}C^{-1},\ B^{-1}C^{-1}A \to AC^{-1}B^{-1}, \\ & C^{-1}AB \to BAC^{-1},\ \ B(CA^{-1})^nB^{-1} \to (A^{-1}C)^n, \\ & A(BC)^nA^{-1} \to (CB)^n,\ B(AC^{-1})^nB^{-1} \to (C^{-1}A)^n, \\ & C(A^{-1}B^{-1})^nC^{-1} \to (B^{-1}A^{-1})^n\ \ |\ \ n\in\mathbb{N}\ \}. \end{aligned}$$

It is easily seen that the limit systems for all weight functions ω_{ACB}, ω_{BAC}, ω_{BCA}, ω_{CAB}, ω_{CBA} are also infinite.

Additional remarks:

For practical purposes it is important to have algorithms which compute the complete system R^{∞} (if it is finite) fast. An interesting way to avoid the investigation of unnecessary critical pairs is described in [Kü 86]. It should be added, however, that such methods do not always improve the speed of the algorithm : One always has to test whether the method applies and these tests also consume some time and space. Other heuristics and strategies are e.g. first to overlap short words and similar devices; each implementation usually contains such methods. They all have positive as well as negative aspects; theoretical results in this direction are nearly unknown.

II.5. The Special Word Problem for Groups and Small Cancellation Theory

The oldest and most popular algorithm which solves the word problem for certain classes of groups is Dehn's algorithm. It allows to replace u by v if $uv^{-1} = e$ is a defining relator and $|u| > |v|$ for the length function $|\ldots|$. The power of this algorithm was investigated over several decades; a good overview can be found in the book by Lyndon and Schupp on "Combinatorial Group Theory" ([Ly-Sch]).
The most important notion in this context is the concept of small cancellation conditions which are sufficient for the success of Dehn's algorithm.
Small cancellation conditions we already met are the so called $C'(\lambda)$-conditions, $\lambda \in \mathbb{R}^+$.
Remember that $S \subseteq (\Sigma \cup \Sigma^{-1})^*$ satisfies the $C'(\lambda)$-condition iff $wu, wv \in S$ implies $|w| < \lambda \cdot |wv|$.
In this section we need some more conditions, the so called $T(p)$-conditions.

1. Definition:

Let $S \subseteq (\Sigma \cup \Sigma^{-1})^*$ and $p \in \mathbb{N}$ be given.
S satifies the T(p)-condition, iff for all n, $3 \leq n < p$ and $w_1, \ldots, w_n \in S$ where w_i is not the literally inverse of w_{i+1} for $1 \leq i \leq n$, there is at least one of the products $w_1 w_2$, $w_2 w_3, \ldots, w_n w_1$ which is freely reduced.

The following theorem (of Greendlinger) is central for Dehn's algorithm and proved in [Ly-Sch].

2. Theorem:

Suppose $G = \langle \Sigma = \{a_1,\ldots,a_n\} \mid \mathbf{S}\rangle$ is a finitely presented group, **S** symmetrized (i.e. all elements of **S** are cyclically reduced and all cyclic permutations of s or s^{-1} are in **S** for any $s \in \mathbf{S}$) and **S** satisfies $C'(^1/_6)$ or **S** satisfies $C'(^1/_4)$ and $T(4)$.
Then every freely reduced word $w \in (\Sigma \cup \Sigma^{-1})^*$ for which $w = e$ holds in G contains more than half of some element of **S**. These conditions are sharp among the conditions of type $C'(\lambda)$ and $T(p)$.

As demonstrated in section II.3. the Knuth-Bendix completion algorithm, in particular the group completion algorithm GCA, generates reduction systems namely Sym(**R**) which contain Dehn's algorithm as a subsystem. This means that all sufficient conditions for Dehn's algorithm are also sufficient for the GCA. But there it has some disadvantage simply to take over the combinatorial results of the small cancellation theory: One does not get an insight into the procedures and does not get any information on the impact of stronger algorithms.
For this reason new sufficient conditions for reduction systems were developed by Hans Bücken (cf. [Bü79]). These conditions are of non-numerical kind but some of them imply the small cancellation conditions. Hence the following section (II.5.1.) contains most of the results on Dehn's algorithm as a special case, e.g. Greendlinger's results on the $C'(^1/_6)$-groups.

II.5.1. Superpos - Deduction - Chains and Criteria for the Solvability of the Word Problem

In this section we denote a finitely presented group $G_W(\Sigma/\mathbf{R})$ also by $\langle a_1,\dots,a_n|\mathbf{S}\rangle$, where $\mathbf{S} = \{lr^{-1}|\ l\to r \in \mathbf{R}\}$, and $\Sigma = \{a_1,\dots,a_n\}$. Furthermore we assume that

$$\mathrm{Sym}(\mathbf{S}) := \{t|\ t \text{ is a cyclic permutation of } s \text{ or } s^{-1} \text{ for some } s\in\mathbf{S}\}$$

satisfies $C'(^1/_2)$.

Since **R** is supposed to be irreducible with respect to **FR**, every element of Sym(**S**) is freely reduced.

Furthermore we fix throughout our discussion a G-ordering $\ll$ on $(\Sigma \cup \Sigma^{-1})^*$ by the weight function ω where $\omega(a_i) = 1$ for $1\le i\le n$. Hence $\ll$ is just the ordering which can be obtained by ordering words according to their length $|\dots|$ and lexicographically if they have the same length.

First we study the effect of Sym(**R**) $\cup$ **FR** =: $\mathbf{R}_G$ where Sym(**R**) is generated with respect to $\ll$.

3. Lemma:

$|\mathrm{Irred}(t,\mathbf{R}_G)| = 1$ for every subword t of some $s \in \mathrm{Sym}(\mathbf{S})$.

Proof: We take $t = l_1l_2l_3l_4l_5$ such that $l_2l_3 \to r_1,\ l_3l_4 \to r_2 \in \mathbf{R}_G$.

Then $C'(^1/_2)$ implies $l_3r_1^{-1}l_2 = l_3l_4r_2^{-1} = l_3l_4l_5vl_1l_2$.

Therefore we obtain: $r_1^{-1} = l_4 l_5 v l_1,\ r_2^{-1} = l_5 v l_1 l_2$

and

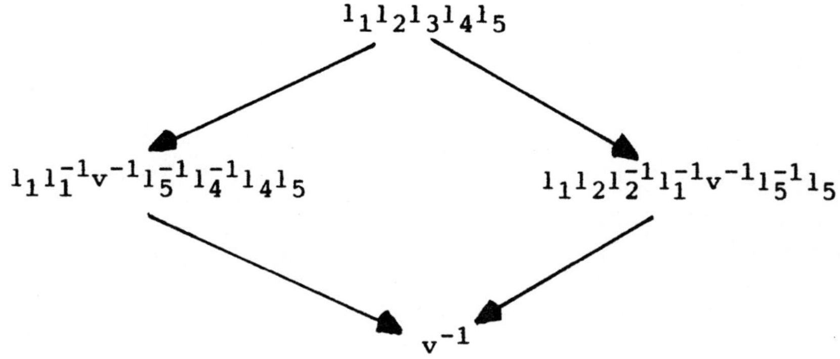

[]

As an immediate consequence we remark:

Suppose Sym(**S**) satisfies $C'(^1/_2)$ and (w_1,w_2) is a critical pair of Sym(**R**) then there is $l_1 u \to r_1,\ u l_2 \to r_2 \in$ Sym(**R**) such that $w_1 = r_1 l_2$ and $w_2 = l_1 r_2$, in particular, there is no critical pair $(r_1, u r_2 v)$ such that $u l_2 v \to r_1,\ l_2 \to r_2 \in$ Sym(**R**).

Our next aim is to describe a kind of a "standard deduction" which we will call superpos-deduction-chain.

The idea runs as follows:
Starting at an element of a critical pair, this word is reduced as far as possible. Now at the left or at the right side of the irreducible word just as much is added that another reduction is applicable, and so on. Then the same procedure is applied to the other word of the critical pair.

More formally:

4. Definition:

Let (w_1,w_2) be a critical pair of Sym(**R**).
A sequence $(E_n,I_n,\ldots,E_o,I_o,w_1,w_2)$ is called a <u>superpos-deduction-chain of length n</u> if:

(1) If $n = 0$ then $E_o = w_1$, $I_o = \mathrm{Irred}(w_1,\mathbf{R}_G)$
(2) If $n > 0$ then each E_k, $0<k\leq n$ is of the form
$E_k = s_p\ldots s_1 I_o t_1\ldots t_q$, $p+q=k$
subject to the following conditions:

either (a) $E_k = s_p E_{k-1}$ and
(b) $s_p\ldots s_1$ is irreducible with respect to $\mathbf{R}_G$
(c) $s_p I_{k-1}$ is reducible with respect to $\mathbf{R}_G$
(d) $I_k = \mathrm{Irred}(s_p I_{k-1},\mathbf{R}_G)$

or (α) $E_k = E_{k-1} t_q$
(β) $t_1\ldots t_q$ is irreducible with respect to $\mathbf{R}_G$
(γ) $I_{k-1} t_q$ is reducible with respect to $\mathbf{R}_G$
(δ) $I_k = \mathrm{Irred}(I_{k-1} t_q,\mathbf{R}_G)$

If $E_n = sI_o t$ for $n>0$, we put $E_n(w_2) := s w_2 t$; in particular we put $E_o(w_2) := w_2$.

<u>5. Example:</u> Let $\langle\Sigma/\mathbf{S}\rangle = G_w(\Sigma/\mathbf{R})$ be a finitely presented group.

Assume $l_k u_k \to r_k \in \mathrm{Sym}(\mathbf{R})$ for $k\in\{0,1,2\}$,
$u_i l_i \to r_i \in \mathrm{Sym}(\mathbf{R})$ for $i\in\{4,5\}$
and $u_o l_3 \to r_3 \in \mathrm{Sym}(\mathbf{R})$.

Then we have for instance the following superpos-deduction-chain of length 4 for $(w_1,w_2) = (r_o l_3, l_o r_3)$

$l_2 l_1 z^{-1} l_o u_o l_3 l_4 l_5$ $\qquad$ $w_2 = l_o r_3 = E_o(w_2)$

$E_o = w_1 = r_o l_3$ $\qquad$ $l_2 l_1 z^{-1} r_o l_3 l_4 l_5$ $\qquad$ $l_2 l_1 z^{-1} l_o r_3 l_4 l_5 = E_4(w_2)$

$\downarrow *$

$I_o = z u_1 v_o$ $\qquad$ $l_2 l_1 z^{-1} z u_1 v_o l_4 l_5$

$\downarrow *$

$E_1 = l_1 z^{-1} I_o$ $\qquad$ $l_2 r_1 v_o l_4 l_5$ $\qquad$ $E_1(w_2) = l_1 z^{-1} w_2$

$\downarrow *$

$I_1 = v_1 u_4$ $\qquad$ $l_2 v_1 u_4 l_4 l_5$ $\qquad$ $E_2(w_2) = l_1 z^{-1} w_2 l_4$

$\downarrow$

$E_2 = l_1 z^{-1} I_o l_4$ $\qquad$ $l_2 v_1 r_4 l_5$

$\downarrow *$

$I_2 = v_2 u_5$ $\qquad$ $l_2 v_2 u_5 l_5$ $\qquad$ $E_3(w_2) = l_1 z^{-1} w_2 l_4 l_5$

$\downarrow$

$E_3 = l_1 z^{-1} I_o l_4 l_5$ $\qquad$ $l_2 v_2 r_5$

$\downarrow *$

$I_3 = u_2 v_3$ $\qquad$ $l_2 u_2 v_3$ $\qquad$ $E_4(w_2) = l_2 l_1 z^{-1} w_2 l_4 l_5$

$\downarrow$

$E_4 = l_2 l_1 z^{-1} I_o l_4 l_5$ $\qquad$ $r_2 v_3$

$\downarrow *$

$I_4 = v_4$ $\qquad$ v_4

Next we will obtain a first (more technical) sufficient condition for the solvability of the word problem.

6. Theorem:

Suppose $G = \langle a_1,\dots,a_n \mid \mathbf{S}\rangle$ $(=G(\Sigma/\mathbf{R}),\ \mathbf{R}=\{s\to e \mid s\in\mathbf{S}\})$ and $\mathrm{Sym}(\mathbf{S})$ satisfies $C'(^1/_2)$.
For all critical pairs (w_1,w_2) of $\mathrm{Sym}(\mathbf{R})$ and all superpos-deduction-chains $(E_m,I_m,\dots,E_o,I_o,w_1,w_2)$, $m\geq 0$, such that $L(E_m,\mathbf{R}_G) \cap L(E_m(w_2),\mathbf{R}_G) = \emptyset$ for $0 \leq k \leq m$ assume $e \in \mathrm{Irred}(I_m^{-1}E_m(w_2),\mathbf{R}_G)$.

Then $u \underset{\mathbf{R}_G}{\leftrightarrow^*} e$ implies $\mathrm{Irred}(u,\mathbf{R}_G) = \{e\}$.

Proof: Let $u \underset{\mathbf{R}_G}{\leftrightarrow^*} e$ be freely reduced.

There is some word $v \underset{\mathbf{FR}}{\leftrightarrow^*} u$ such that

$$v = t_1^{-1}s_1t_1t_2^{-1}s_2t_2\dots t_k^{-1}s_kt_k, \quad s_i\in S \quad \text{for} \quad 1\leq i\leq k.$$

Obviously v can be reduced to e by $\mathbf{R}_G$. Since $u \in L(v,\mathbf{R}_G)$ it suffices to prove $\mathrm{Irred}(v,\mathbf{R}_G) = \{e\}$.

We proceed indirectly and assume $\mathrm{Irred}(v,\mathbf{R}_G) \supset \{e\}$.

We put $w = \min\{t \mid t\in L(v,\mathbf{R}_G),\ t \underset{\mathbf{R}_G}{\to} w_1,\ t \underset{\mathbf{R}_G}{\to} w_2,$
$\mathrm{Irred}(w_1,\mathbf{R}_G)=\{e\}$ and $e\notin \mathrm{Irred}(w_2,\mathbf{R}_G)\}$

Because of the minimality of w there exist $l_1a\to r_1$, $al_2\to r_2$ in $\mathrm{Sym}(\mathbf{R})$ such that

$$w = v_1l_1al_2v_2$$
$$w_1 = v_1r_1l_2v_2$$
$$w_2 = v_1l_1r_2v_2$$

with $\mathrm{Irred}(v_i,\mathbf{R}_G) = v_i$ for $i=1,2$.

Because $Irred(w_1,R_G) = \{e\}$ there is a superpos-deduction-chain $(E_m,I_m,\dots,E_1,I_1,E_o,I_o,r_1l_2,l_1r_2),m>0$, which describes the reduction of w_1 to e where $E_m = v_1 Irred(l_1r_2,R_G)v_2$ and $I_m = e$.
From the definition of w we get:

$$L(r_1l_2,R_G) \cap L(l_1r_2,R_G) = \emptyset$$

and $$L(E_k,R_G) \cap L(E_k(l_1r_2),R_G) = \emptyset \quad \text{for} \quad 0<k\le m.$$

The hypothesis of the theorem implies:

$$e \in Irred(I_m^{-1}E_m(l_1r_2),R_G)$$

which yields:

$$e \in Irred(e^{-1}E_m(l_1r_2),R_G)$$
$$e \in Irred(E_m(l_1r_2),R_G)$$
$$e \in Irred(w_2,R_G) \;;$$

a contradiction to the assumption.

[]

The disadvantage of the sufficient condition in the last theorem lies in its non-constructive nature. In order to obtain a more algorithmic version we have in particular to restrict the conditions (d) and (δ) in the definition of a superpos-deduction-chain.

7. Definition:

R_G satisfies the condition K1 iff for some critical pair (w_1,w_2) of Sym(R) $Irred(w_i,R_G) \neq Irred(w_i,FR)$ for i=1,2 implies $Irred(w_1,R_G) = Irred(w_2,R_G)$.

Obviously we have: If **R** is complete, then **R** satisfies K1. On the other hand if **R** satisfies K1 the CR-property can be violated only if

$$\text{Irred}(w_i,\mathbf{R}\cup\mathbf{FR}) = \text{Irred}(w_i,\mathbf{FR}).$$

Hence if **R** satisfies K1 and $(E_m,I_m,\ldots,E_o,I_o,w_1,w_2)$ is a superpos-deduction-chain for some critical pair (w_1,w_2) of **R** satisfying $L(w_1,\mathbf{R}) \cap L(w_2,\mathbf{R}) = \emptyset$ then

$$I_o = \text{Irred}(w_1,\mathbf{R}) = \text{Irred}(w_1,\mathbf{FR}).$$

The following lemmas are useful for the further discussion.

8. <u>Lemma:</u>

Suppose $\mathbf{R}_G$ satisfies K1 and let (w_1,w_2) be a critical pair occuring from $L_1 \to R_1$, $L_2 \to R_2 \in \text{Sym}(\mathbf{R})$ such that $\text{Irred}(w_i,\mathbf{R}_G)=\text{Irred}(w_i,\mathbf{FR})$ and $\text{Irred}(w_1,\mathbf{FR}) \neq \text{Irred}(w_2,\mathbf{FR})$ for $i=1,2$.

Then we present:

$$L_1 = l_1v_1^{-1}u \qquad R_1 = r_1v_2$$
$$L_2 = uv_2^{-1}l_2 \qquad R_2 = v_1r_2$$

with $\text{Irred}(w_1,\mathbf{R}_G)=r_1l_2$, $\text{Irred}(w_2,\mathbf{R}_G)=l_1r_2$, $r_1^{-1}l_1 \neq l_2r_2^{-1}$.

<u>Proof:</u> We look at

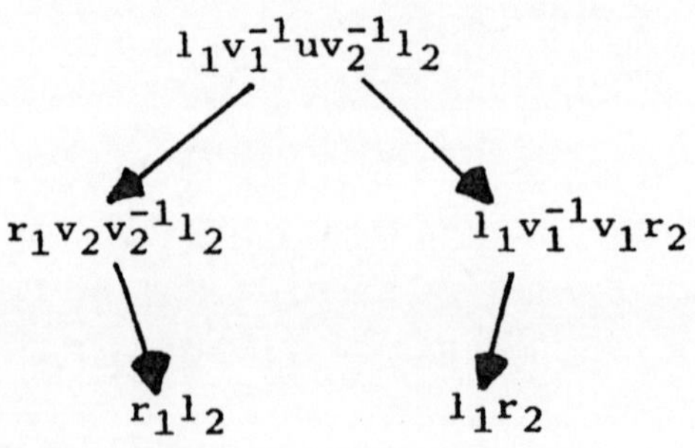

and assume $r_1^{-1}l_1 = l_2r_2^{-1}$.

Then we have: either a) $r_1^{-1} = l_2t$ and $r_2^{-1} = tl_1$
or b) $l_2 = r_1^{-1}t$ and $l_1 = tr_2^{-1}$

Case a) $r_1^{-1} = l_2t$ implies $r_1 = t^{-1}l_2^{-1}$, $r_2 = l_1^{-1}t^{-1}$

From $r_1l_2 = t^{-1}l_2^{-1}l_2 \rightarrow t^{-1}$
and $l_1r_2 = l_1l_1^{-1}t^{-1} \rightarrow t^{-1}$
we get a contradiction.

Case b) is handled in the same manner.

[]

9. Lemma:

If for $(lu,vr) \in Sym(\mathbf{R})$ $v^{-1}l$ is irreducible with respect to $\mathbf{R}_G$, we have $|v| \le |u|$.

Proof: Because $v^{-1}l$ is irreducible one obviously obtains:

(1) $|lu| = |l|+|u| \ge |v|+|r| = |vr|$

(2) $|v^{-1}l| = |v|+|l| \le |r|+|u| = |ru^{-1}|$

Hence we have: $|l|+|u|-(|v|+|l|) \ge |v|+|r|-(|r|+|u|)$,

which implies $|v| \le |u|$.

[]

10. Definition:

Suppose R_G satisfies K1 and let (w_1,w_2) be a critical pair of Sym(R) which violates the CR-property, such that in the notation of Lemma 8 we have:

$$\mathrm{Irred}(w_1,R_G) = r_1 l_2 \neq l_1 r_2 = \mathrm{Irred}(w_2,R_G)$$

Further assume $D \in \{D_l, D_r\}$, where $r_1 = r_{1l} D_l$ and $l_2 = D_r l_{2r}$.

(1) D is called a <u>block</u> for (w_1,w_2) iff there is a superpos-deduction-chain $(E_1,I_1,w_1,I_0,w_1,w_2)$ such that either

(i) $D = D_l$ (D_l is also called a <u>left block</u>)

(ii) $E_1 = s r_1 l_2$

(iii) $L(E_1,R_G) \cap L(s l_1 r_2, R_G) = \emptyset$

(iv) $I_1 = \mathrm{Irred}(s r_{1l}, R_G) D_l l_2$

or

(i) $D = D_r$ (D_r is also called a <u>right block</u>)

(ii) $E_1 = r_1 l_2 t$

(iii) $L(E_1,R_G) \cap L(l_1 r_2 t, R_G) = \emptyset$

(iv) $I_1 = r_1 D_r \, \mathrm{Irred}(l_{2r} t, R_G)$

(2) <u>R_G satisfies K2</u> iff there is a block D for every critical pair which violates the CR-property.

(3) Let D_l be a left block and D_r be a right block for the critical pair (w_1,w_2). $D_l D_r$ is called <u>invincible</u> if

(i) $D_l D_r$ is irreducible with respect to R_G

(ii) For $D_l = D_l'a$, aD_r is not a subword of some $l \in \mathrm{Left(Sym(R))}$

(iii) For $D_r = aD_r'$, $D_l a$ is not a subword of some $l \in \mathrm{Left(Sym(R))}$.

(4) <u>R_G satisfies K</u> iff R_G satisfies K1 and K2 and if all left blocks D_l resp. all right blocks D_r are invincible.

Now we see at once:

11. Lemma:

Suppose $w = D_1D_2...D_k$, D_i is a left block for (w_i,w_{i+1}), $1\le i\le n\le k$, D_j is a right block for (w_{j-1},w_j), $n+1\le j\le k$, and D_mD_{m+1} is invincible for $1\le m<k$, then w is irreducible.

12. Lemma:

Let $\mathbf{R}_G$ satisfy condition K and suppose $(E_n,I_n,...,E_1,I_1,r_oL_o,I_o,r_oL_o,l_oR_o)$ is a superpos-deduction-chain of length n, where $E_n = s_p...s_1I_ot_1...t_q$ and $I_k \notin \mathrm{Irrred}(E_k(w_2),\mathbf{R}_G)$ for $1\le k\le n$.
Then $I_n = S_pD^l_{p-1}...D^l_1D^l_oD^r_o...D^r_{q-1}T_q$, $n=p+q$, where D^l_i,D^r_j are blocks for $0\le i\le p-1$ and $0\le j\le q-1$ and the products $D^l_iD^l_{i-1}$, $D^l_oD^r_o$ and $D^r_jD^r_{j+1}$ are invincible for $0<i\le p-1$ and $0\le j<q-1$.

Proof: The proof by induction on the length of the superpos-deduction-chain is left to the reader.

[]

Now we can replace the hypothesis of Theorem 6 by the condition K.

13. Theorem:

Let $\mathbf{R}_G$ have property K and assume $(E_m,I_m,...,E_1,I_1,w_1,I_o,w_1,w_2)$ is a superpos-deduction-chain for some critical pair (w_1,w_2) of $\mathrm{Sym}(\mathbf{R})$ such that $I_m \notin \mathrm{Irred}(E_m(w_2),\mathbf{R}_G)$ and $L(w_1,\mathbf{R}_G) \cap L(w_2,\mathbf{R}_G) = \emptyset$.
Then $e \in \mathrm{Irred}(I_m^{-1}E_m(w_2),\mathbf{R}_G)$.

<u>Proof:</u> In the notation of Lemma 12 we have

$$I_m = S_p D^l_{p-1} \ldots D^l_o D^r_o \ldots D^r_{q-1} T_q, \qquad m=p+q.$$

The proof is by induction on m.

First we consider: $m = 0$.

Since $\mathbf{R}_G$ satisfies in particular condition K1 there are according to Lemma 8 $l_1 u \to r_1$, $u l_2 \to r_2 \in \mathrm{Sym}(\mathbf{R})$ such that $r_1 l_2 = w_1 = \mathrm{Irred}(w_1, \mathbf{R}_G)$ and $l_1 r_2 = w_2 = \mathrm{Irred}(w_2, \mathbf{R}_G)$ whenever (w_1, w_2) is a critical pair which violates the CR-property.

Hence we have $I_o^{-1} E_o(w_2) = w_1^{-1} w_2 = l_2^{-1} r_1^{-1} l_1 r_2$.

Assuming $r_1^{-1} l_1$ is irreducible we obtain $u^{-1} \to r_1^{-1} l_1$, i.e. $|u| \geq {}^1/_2 |r_1 l_1|$ which contradicts the fact that $\mathrm{Sym}(\mathbf{S})$ satisfies $C'({}^1/_2)$. Therefore $r_1^{-1} l_1$ is reducible to u^{-1} (by Lemma 3) and hence $l_2^{-1} r_1^{-1} l_1 r_2 \to l_2^{-1} u^{-1} r_2 \to^* e$, what we wanted to show.

For the induction step assume $e \in \mathrm{Irred}(I_{m-1}^{-1} E_{m-1}(w_2), \mathbf{R}_G)$.

Again we have two cases:

(a) $E_m(w_2) = s_p E_{m-1}(w_2)$.

Then we have:

$$I_m^{-1} E_m(w_2) = T_q^{-1} (D^r_{q-1})^{-1} \ldots (D^l_{p-1})^{-1} S_p^{-1} s_p E_{m-1}(w_2)$$

As a consequence of Lemma 12 we have:

$$s_p = l_p v_{p-1}^{-1} \text{ and } l_{p-1} u_{p-1} \to v_{p-1} w_{p-1} D^l_{p-1} \in \mathrm{Sym}(\mathbf{R})$$
$$l_p w_{p-1} \to S_p \in \mathrm{Sym}(\mathbf{R})$$

This and $C'({}^1/_2)$ implies $S_p^{-1} l_p \to w_{p-1}^{-1}$.

Thus we have:

$$I_m^{-1} E_m(w_2) = T_q^{-1} (D^r_{q-1})^{-1} \ldots (D^l_{p-1})^{-1} S_p^{-1} l_p v_{p-1}^{-1} E_{m-1}(w_2)$$

$$\to T_q^{-1} (D^r_{q-1})^{-1} \ldots (D^l_{p-1})^{-1} w_{p-1}^{-1} v_{p-1}^{-1} E_{m-1}(w_2)$$

$$= I_{m-1}^{-1} E_{m-1}(w_2)$$

The induction hypothesis implies

$$e \in \mathrm{Irred}(I_{m-1}^{-1}E_{m-1}(w_2), \mathbf{R}_G).$$

(b) $E_m(w_2) = E_{m-1}(w_2)S_q$.
The proof is as in case (a).

[]

14. Corollary:

If $\mathbf{R}_G$ satisfies K and $C'(^1/_2)$ then the reduction system $\mathbf{R}_G$ solves the word problem.

Proof: Theorem 6 and Theorem 13.

The property K can be tested effectively although it is not of numerical character like the small cancellation conditions.

II.5.2. The Small Cancellation Conditions and the Condition K

This section establishes the connection between the condition K and the small cancellation conditions. We derive in particular some classical results for the solvability of the word problem. The notation is the same as in the last section; the next Lemma is connected with Lemma 8.

15. Lemma:

Suppose R_G satisfies $C'(\lambda)$ with $\lambda \le {}^1/_4$.
Let $L_1 = l_1v_1u$, $L_2 = uv_2l_2$, $R_1 = r_1v_2^{-1}$, $R_2 = v_1^{-1}r_2$, $w_1 = r_1v_2^{-1}v_2l_2$, $w_2 = l_1v_1v_1^{-1}r_2$ and assume

$$|L_1R_1^{-1}| \ge |L_2R_2^{-1}|.$$

Then the following holds:

(1) If $|v_1uv_2| \ge \lambda|L_1R_1^{-1}|$ then $\mathrm{Irred}(w_1,R_G) = \mathrm{Irred}(w_2,R_G)$.

(2) If $|v_1uv_2| < \lambda|L_1R_1^{-1}|$ and $r_1^{-1}l_1 = l_2r_2^{-1}$ then $\mathrm{Irred}(w_1,R_G) = \mathrm{Irred}(w_2,R_G)$.

(3) If $|v_1uv_2| \le \lambda|L_2R_2^{-1}|$ and $r_1l_1 \neq l_2r_2^{-1}$ then r_1l_2 and l_1r_2 are irreducible with respect to R_G.

Proof: Concerning (1) and (2) we observe $r_1^{-1}l_1 = l_2r_2^{-1}$. Therefore we have either

a) $l_2 = r_1^{-1}t$ and $l_1 = tr_2^{-1}$ or

b) $r_1^{-1} = l_2t$ and $r_2^{-1} = tl_1$

The only reductions are:

Case a) : $\text{Irred}(w_1,\mathbf{FR}) = r_1l_2 = r_1r_1^{-1}t \to t$

$\text{Irred}(w_2,\mathbf{FR}) = l_1r_2 = tr_2^{-1}r_2 \to t$

Case b) : $\text{Irred}(w_1,\mathbf{FR}) = r_1l_2 = t^{-1}l_2^{-1}l_2 \to t^{-1}$

$\text{Irred}(w_2,\mathbf{FR}) = l_1r_2 = l_1l_1^{-1}t^{-1} \to t^{-1}$

In order to prove (3) we consider only one case and assume that some $(l,r) \in \text{Sym}(\mathbf{R})$ reduces r_1l_2.
If we write $r_1 = r_1'l'$ and $l_2 = l''l_2''$, then we have two possibilities:

(a) $|l'| \geq {}^1/_4|lr^{-1}| \geq \lambda|lr^{-1}|$ or

(b) $|l''| \geq {}^1/_4|lr^{-1}| \geq \lambda|lr^{-1}|$.

Both cases are handled in the same way and we restrict ourselves to (a). We have $l''r^{-1} = v_2^{-1}L_1^{-1}r_1'$, but $v_2^{-1} = e$ because L_2 is freely reduced.
Hence $l''r^{-1} = L_1^{-1}r_1' = u^{-1}v_1^{-1}l_1^{-1}r_1'$.
Therefore L_2 is not freely reduced, a contradiction.

[]

16. Lemma:

Assume $\mathbf{R}_G$ satisfies K1
and $l_1v_1u \to r_1v_2^{-1},\ uv_2l_2 \to v_1^{-1}r_2 \in \mathbf{R}_G$ and $r_1 = zr_{1r}$.
Then $r_{1r}l_2 \ast \text{Irred}(z^{-1}l_1r_2,\mathbf{R}_G)$ implies $z^{-1}l_1$ is irreducible, i.e. $r_{1r}v_2^{-1}u^{-1}v_1^{-1} \to z^{-1}l_1$.

Proof: Suppose $z^{-1}l_1$ is reducible.
From $C'(^1/_2)$ and Lemma 3 we get:

$$z^{-1}l_1 \to r_{1r}v_2^{-1}u^{-1}v_1^{-1}$$

and

$$r_{1r}v_2^{-1}u^{-1}v_1^{-1}r_2 \to r_{1r}l_2$$

which is a contradiction.

[]

17. Theorem:

Suppose $G = \langle a_1,\dots,a_n | S\rangle$. If $Sym(S)$ satisfies $C'(^1/_6)$ then R_G has the property K.

Proof: By Lemma 15 R_G satisfies K1. So according to Lemma 8 we have for every critical pair (w_1,w_2) violating the CR-property:

There are $L_1 \to R_1$, $L_2 \to R_2 \in Sym(R)$ such that
$L_1 = l_1v_1u$, $L_2 = uv_2l_2$, $R_1 = r_1v_2^{-1}$, $R_2 = v_1^{-1}r_2$
with $w_1 = r_1v_2^{-1}v_2l_2$ and $w_2 = l_1v_1v_1^{-1}r_2$,
$Irred(w_1,R_G) = r_1l_2$ and $Irred(w_2,R_G) = l_1r_2$.

In order to show K2 it is sufficient to consider the cases

A) $E_1 = l_3z^{-1}zv_3v_4^{-1}D_1l_2$, $r_1 = zv_3v_4^{-1}D_1$, $l_3v_3 \to r_3v_4 \in R_G$

B) $E_1 = r_1D_rv_4^{-1}v_3zz^{-1}l_3$, $l_2 = D_rv_4^{-1}v_3z$, $v_3l_3 \to v_4r_3 \in R_G$.

Case A) We divide the proof into parts:

a) $|v_3v_4^{-1}| < {^1/_6}\min(|l_3v_3v_4^{-1}r_3^{-1}|, |L_1R_1^{-1}|)$

Assuming the contrary $C'(^1/_6)$ yields:

$$v_3v_4^{-1}r_3^{-1}l_3 = v_3v_4^{-1}D_1v_2^{-1}u^{-1}v_1^{-1}l_1^{-1}z$$

and we have $z = e$ because $l_3 z^{-1}$ is irreducible.
W.l.o.g. we assume $l_3 = v_1^{-1} l_1^{-1}$.
(the cases $l_3 = u_r^{-1} v_1^{-1} l_1^{-1}$, where $u = u_l u_r$,
$l_3 = v_{r1}^{-1} l_1^{-1}$, where $v_1 = v_{1l} v_{1r}$
and $l_3 = l_{1r}^{-1}$, where $l_1 = l_{1l} l_{1r}$
can be treated in the same way.)

Then we obtain:

$$\begin{aligned} E_1 &= l_3 v_3 v_4^{-1} D_1 l_2 && \to r_3 v_4 v_4^{-1} D_1 l_2 \\ &= u v_2 D_1^{-1} v_4 v_4^{-1} D_1 l_2 && \to u v_2 l_2 \\ & && \to v_1^{-1} r_2 \end{aligned}$$

and $$l_3 l_1 r_2 = v_1^{-1} l_1^{-1} l_1 r_2 \to v_1^{-1} r_2.$$

This is a contradiction to

$$L(E_1, R_G) \cap L(s l_1 r_2, R_G) = \emptyset.$$

b) $|D_1| \geq {}^1/_6 |L_1 R_1^{-1}|$

Again we proceed indirectly.
First of all we have according to Lemma 16

$$t := v_3 v_4^{-1} D_1 v_2^{-1} u^{-1} v_1^{-1} = r_{1r} v_2^{-1} u^{-1} v_1^{-1} \to z^{-1} l_1$$

Putting together $|v_3 v_4^{-1}| < {}^1/_6 |L_1 R_1^{-1}|$,
$|D_1| < {}^1/_6 |L_1 R_1^{-1}|$ and
$|v_2^{-1} u v_1^{-1}| < {}^1/_6 |L_1 R_1^{-1}|$
we obtain

$$|t| < {}^1/_2 \, |L_1 R_1^{-1}|.$$

Therefore t is irreducible; this contradiction shows (b).

Next we consider $E_1 = l_3 z^{-1} z v_3 v_4^{-1} D_1 l_2 \to r_3 v_4 v_4^{-1} D_1 l_2$
$\to r_3 d_1 l_2$.

c) $r_3 D_1 l_2$ is irreducible with respect to R_G.

Assuming that this is not the case there is $L \to R \in R_G$ such that either

(i) D_1 is a subword of L, i.e. $L = r_{3r} D_1 l_{21}$

or (ii) $L = r_{3r} d_1$, where $D_1 = D_{1l} D_{1r}$, $r_3 = r_{3l} r_{3r}$ and $l_2 = l_{21} l_{2r}$.

In both cases we have:

(α) $|D_1| \geq {}^1/_6 |L_1 R_1^{-1}|$
or (β) $|r_{3r}| \geq {}^1/_6 |LR^{-1}|$
or (γ) $|l_{21}| \geq {}^1/_6 |LR^{-1}|$

<u>Case (i):</u> By $C'({}^1/_6)$ we have in case (α):

$$D_1 v_2^{-1} u^{-1} v_1^{-1} l_1^{-1} z v_3 v_4^{-1} = D_1 l_r R^{-1} l_l = D_1 l_{21} R^{-1} r_{3r}$$

$v_3 v_4^{-1} = e$ is impossible ($v_3 \neq e$ by construction), hence $l_l = r_{3r} = e$ because $l_3 v_3 v_4^{-1} r_3^{-1}$ is freely reduced; and in the same way we get $l_r = l_{21} = e$. Thus D_1 is reducible which contradicts the fact that $r_1 = z v_3 v_4^{-1} D_1$ is irreducible.

In case (β) we have by $C'({}^1/_6)$:

$r_{3r} D_1 l_{21} R^{-1} = r_{3r} v_4 v_3^{-1} l_3^{-1} r_{31}$, which contradicts the fact that $r_1 = z v_3 v_4{}^{-1} d_1$ is freely reduced.

Finally in case (γ) $C'(^1/_6)$ yields:

$$l_{21}R^{-1}r_{3r}D_1 = l_{21}l_{2r}r_2^{-1}v_1uv_2.$$

Hence v_2=e because $R_1 = r_1v_2^{-1} = zv_3v_4^{-1}D_1v_2^{-1}$ is freely reduced. But since u$\neq$e by construction there is a contradiction to the fact that
$L_1R_1^{-1} = l_1v_1ur_1^{-1} = l_1v_1uD_1v_4v_3^{-1}z^{-1}$ is freely reduced.

Case (ii) can be shown analogously.

So we have that $r_3D_1l_2$ is irreducible which implies that in case (A) D_1 is a block.

<u>Case B:</u> The arguments are as in case (A).
Thus $\mathbf{R}_G$ satisfies K2.

In order to verify property K consider two blocks D_l and D_r for (w_1,w_2). D_lD_r is irreducible, because r_1l_2 is irreducible.
Suppose $D_l = D_l'a$ and aD_r is a subword of some L$\in$Left($\mathbf{R}_G$), i.e. $L \to R \in \mathbf{R}_G$.
We have $|aD_r| > |D_r| \geq {}^1/_6|L_2R_2^{-1}|$. Put $L = t_1aD_rt_2$; then $C'(^1/_6)$ implies $D_rt_2R^{-1}t_1a = D_rtr_2^{-1}v_1uv_2$.

Therefore v_2 ends with a and so v_2^{-1} starts with a^{-1}.
But $r_1 = sD_l'a$ ends with a ; this contradicts the fact that $R_1 = r_1v_2^{-1}$ is freely reduced.
So v_2 = e and u ends with a. Again we have a contradiction to the fact that $L_1R_1^{-1} = l_1v_1ur_1^{-1}$ is freely reduced.

Analogously it is shown that for $D_r = bD_r'$ D_lb is not a subword of some $L \in$ Left($\mathbf{R}_G$).

[]

18. Theorem:

Suppose $G = \langle a_1,\ldots,a_n \mid S\rangle$. If Sym(S) satisfies $C'(^1/_6)$ then $\mathbf{R}_G$ has the property K.

Proof: By Lemma 15 $\mathbf{R}_G$ satisfies K1. So again according to Lemma 8 we have for every critical pair (w_1,w_2) violating the CR-property:

There are $L_1 \to R_1,\ L_2 \to R_2 \in \mathrm{Sym}(\mathbf{R})$ such that
$L_1 = l_1v_1u,\quad L_2 = uv_2l_2,\quad R_1 = r_1v_2^{-1},\quad R_2 = v_1^{-1}r_2,$
$w_1 = r_1v_2^{-1}v_2l_2,\quad w_2 = l_1v_1v_1^{-1}r_2,$
$\mathrm{Irred}(w_1,\mathbf{R}_G) = r_1l_2$ and $\mathrm{Irred}(w_2,\mathbf{R}_G) = l_1r_2.$

For K2 we have to look (essentially) at two cases:

(A) $E_1=l_3z^{-1}zv_3v_4^{-1}D_1l_2,\quad r_1=zv_3v_4^{-1}D_1,\quad l_3v_3\to r_3v_4 \in \mathbf{R}_G$
(B) $E_1=r_1D_rv_4^{-1}v_3zz^{-1}l_3,\quad l_2=D_rv_4^{-1}v_3z,\quad v_3l_3\to v_4r_3 \in \mathbf{R}_G$

The proof of (B) is omitted, we consider only case (A) where all essential steps occur. We proceed in several steps.

a) $|v_3v_4^{-1}D_1| \geq {}^1/_4|L_1R_1^{-1}|$

Assuming the opposite inequality we have according to Lemma 16

$$v_3v_4^{-1}D_1v_2^{-1}u^{-1}v_1^{-1} = r_{1r}v_2^{-1}u^{-1}v_1^{-1} \to z^{-1}l_1$$
(where $zr_{1r} =: r_1$);

and

$$|v_3v_4^{-1}D_1| + |v_2^{-1}u^{-1}v_1^{-1}| < 1/2|L_1R_1^{-1}|$$

leads to a contradiction.

As in Theorem 17 one obtains:

b) $|v_3v_4^{-1}| < 1/4 \cdot \min(|l_3v_3v_4^{-1}r_3|, |L_1R_1^{-1}|)$

Therefore $D_l \neq e$ and we have for $r_1 = zv_3v_4^{-1}D_l$

$$E_1 = l_3\bar{z}^1 zv_3v_4^{-1}D_l l_2 \rightarrow r_3D_l l_2$$

c) $r_3D_l l_2$ is irreducible with respect to R_G.
Assume the contrary, i.e. $r_3D_l l_2$ is reducible with respect to $L \rightarrow R \in R_G$.

$C'(^1/_4)$ implies $L = t_1D_lt_2$, $t_1,t_2 \neq e$ and
$|t_1| < {}^1/_4 \min(|LR^{-1}|, |L_1R_1^{-1}|)$,
$|t_2| < {}^1/_4 \min(|LR^{-1}|, |l_2R_2^{-1}|)$,
$r_3 = r_{31}t_1$ and $l_2 = t_2l_{2r}$.

Putting $T_1 := D_lt_2R^{-1}t_1 \in \mathrm{Sym}(LR^{-1})$
$T_2 := t_1^{-1}r_{31}^{-1}l_3v_3v_4^{-1} \in \mathrm{Sym}(l_3v_3v_4^{-1}r_3^{-1})$
and $T_3 := v_4v_3^{-1}z^{-1}L_1v_2D_l^{-1} \in \mathrm{Sym}(L_1R_1^{-1})$

we have $T_i \neq T_j$ for $i \neq j$, $i,j \in \{1,2,3\}$ and in addition we have $D_l, t_1, v_3v_4^{-1} \neq e$.
This contradicts condition T(4), i.e. (c) holds.

Therefore D_l is a left block for (w_1,w_2).
Analogously one shows that D_r is a right block for (w_1,w_2).
Hence R_G satisfies K2.

For property K it remains to consider two blocks D_l and D_r. By $C'(^1/_4)$, D_lD_r is irreducible. Assume D_lD_r is not irreducible, i.e. $D_l = D_l'a$ and aD_r is a subword of some $L_o \in \mathrm{Left}(R_G)$, i.e. $L_o = s_1ad_rs_2$.

Putting $S_1 := D_r s_2 R_o^{-1} s_1 a \qquad \in \mathrm{Sym}(L_o R_o^{-1})$

$S_2 := a^{-1}(D_1')^{-1} v_4 v_3^{-1} z^{-1} l_1 v_1 u v_2 \qquad \in \mathrm{Sym}(R_1 L_1^{-1})$

and $S_3 := v_2^{-1} u^{-1} R_2 l_{2r} D_r^{-1} \qquad \in \mathrm{Sym}(R_2 L_2^{-1})$

we have $S_i \neq S_j$ for $i \neq j$, $i,j \in \{1,2,3\}$ which implies $b \neq e$, $uv_2^{-1} \neq e$, $D_r \neq e$ and leads to a contradiction to T(4). Therefore $D_1 D_r$ is invincible.

[]

The last two theorems are slight extensions of classical results on Dehn's algorithm (Dehn's algorithm does not allow reductions (L,R) with $|L| = |R|$). We want to show that also in our case the condition $C'(^1/_6)$ is still sharp (among the $C'(\lambda)$ conditions). To prove it, we take the same group Greendlinger used as a counter example [Gre 60]. The word Greendlinger took as a counterexample is reducible to e with respect to $\mathbf{R}_G$; but another word will do.

19. Example:

$$G = \langle A,B,C \mid ABCA^{-1}B^{-1}C^{-1} \rangle,$$
$$\mathbf{S} = \{ABCA^{-1}B^{-1}C^{-1}\},$$

$$\mathrm{Sym}(\mathbf{S}) = \{ABCA^{-1}B^{-1}C^{-1},\ BCA^{-1}B^{-1}C^{-1}A,\ CA^{-1}B^{-1}C^{-1}AB,\ A^{-1}B^{-1}C^{-1}ABC,\ B^{-1}C^{-1}ABCA^{-1},\ C^{-1}ABCA^{-1}B^{-1},\ CBAC^{-1}B^{-1}A^{-1},\ BAC^{-1}B^{-1}A^{-1}C,\ AC^{-1}B^{-1}A^{-1}CB,\ C^{-1}B^{-1}A^{-1}CBA,\ B^{-1}A^{-1}CBAC^{-1},\ A^{-1}CBAC^{-1}B^{-1}\}$$

$$\mathrm{Sym}(\mathbf{R}) = \{(CBA,\ ABC),\ (A^{-1}CB,\ BCA^{-1}),\ (B^{-1}A^{-1}C,\ CA^{-1}B^{-1}),\ (C^{-1}B^{-1}A^{-1},\ A^{-1}B^{-1}C^{-1}),\ (B^{-1}C^{-1}A,\ AC^{-1}B^{-1}),\ (C^{-1}AB,\ BAC^{-1}),\ (ABCA^{-1},\ CB),\ (BCA^{-1}B^{-1},\ A^{-1}C),\ (CA^{-1}B^{-1}C^{-1},\ B^{-1}A^{-1}),\ (BAC^{-1}B^{-1}, C^{-1}A)\}.$$

G has none of the properties $C'(^1/_6)$, $T(4)$ or K.

The word $w = ABABABCBCA^{-1}CA^{-1}B^{-1}A^{-1}B^{-1}C^{-1}B^{-1}C^{-1}B^{-1}C^{-1}$ is irreducible with respect to $\mathbf{R}_G$, but it is the identity element in G.

Additional Remarks:

The study of Dehn's algorithm goes back to [De 11]. The fact that the relations have common pieces means that critical pairs exist and the small cancellation condition ensures that there is only "little overlapping"; the measures of "little" are taken in a numerical sense like the $C'(^1/_6)$ condition. The number $^1/_6$ arises from the geometrical method of proof and comes ultimately from the Euler polyeder formula.
In the approach presented here, the number $^1/_6$ has a very different origin. Essentially it arises from $3 \cdot {^1/_6} = {^1/_2}$ and the fact that violating the $C'(^1/_2)$ causes immediate difficulties (compare e.g. the proof of Theorem 17). The results of II.5. are essentially contained in Hans Bücken's dissertation [Bü 79].

II.6. Relations between the Completion Procedure and the Todd-Coxeter Algorithm

A famous method to study finitely presented groups is the Todd-Coxeter method (TC in the sequel) of coset enumeration. Suppose $G = \langle \Sigma, S \rangle$ is a finitely presented group and $U \subseteq G$ is a finitely generated subgroup.
The Todd-Coxeter algorithm generates (left resp. right) cosets of U and terminates iff [G:U] is finite.
In the latter case one is also able to construct the coset multiplication table from the output of the algorithm. We restrict ourselves here to the case of the trivial subgroup $U = \{e\}$. Then the algorithm can be interpreted as giving names to group elements and deriving equations between such names. Our aim is to compare the Todd-Coxeter in this light with the completion procedure. For this purpose first a short informal description of the TC-algorithm is given.

We assume that each generator occurs in one of the relators $u = a_1 \ldots a_n$, $a_i \in \Sigma \cup \Sigma^{-1}$. With each such relator a table as shown below is associated:

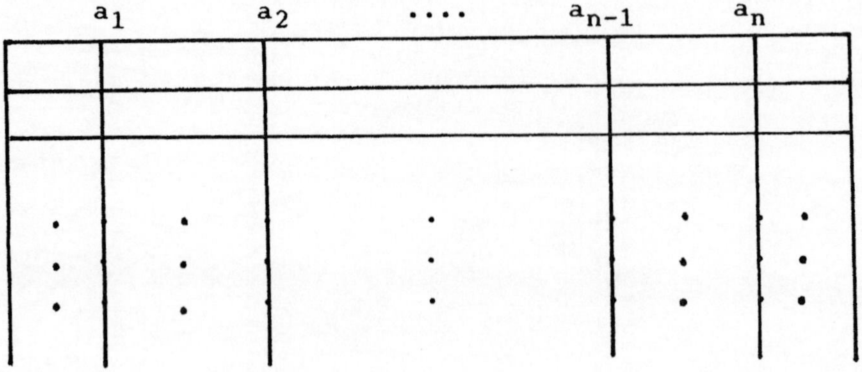

The a_i are markers for the internal dividing lines. The slots will be filled by symbols g_n, $n \in \mathbb{N}$; the g_n are names for (not necessarily different) elements of G.
Insertions in the table are encodings for equations; an entry of the form

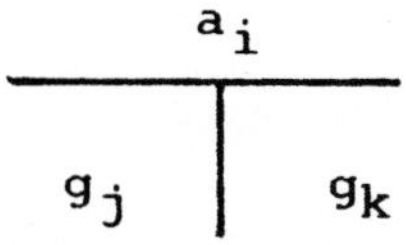

stands for the equations $g_j \cdot a_i = g_k$ and $g_k \cdot a_i^{-1} = g_j$.
Such equations are recorded for convenience in three additional lists: the definition list, the bonus list and the collapse list.
The algorithm as it will be described now is not completely determined because it leaves open how certain tables and slots are selected.

(1) <u>Initial step:</u>

Select some table and write g_1 in the leftmost and rightmost slot of the first row:

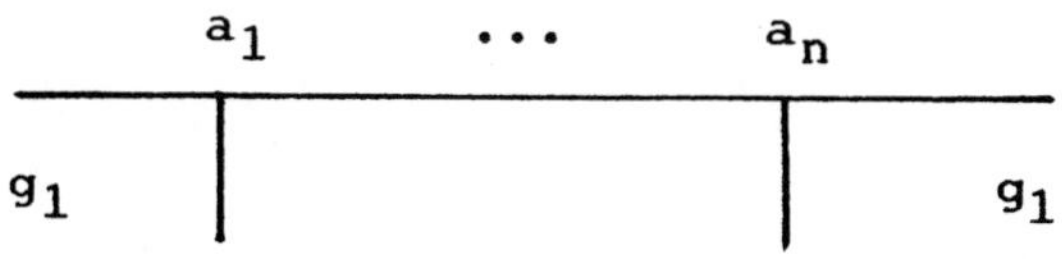

All lists are empty.

(2) <u>Iteration step:</u>

(a) If there is some g_k in the tables and some table such that g_k is not in any row in the leftmost slot than create a new row and write g_k in the leftmost and rightmost slot.

(b) If (a) does not apply then select a row in some table (if it exists) with not all slots filled. Create a new symbol g_k not yet used and put it either in the leftmost or rightmost free slot of the row:

(i) 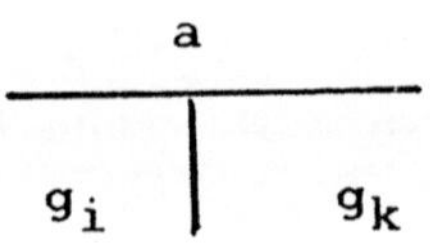

resp. (ii) 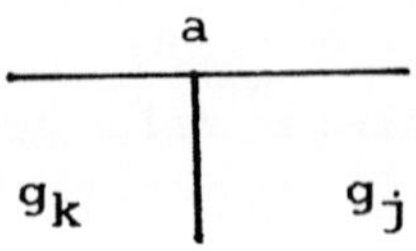

Add the equation $g_i \cdot a = g_k$ resp. $g_j \cdot a^{-1} = g_k$ to the definition list.

If the row is completely filled we have the following situation:

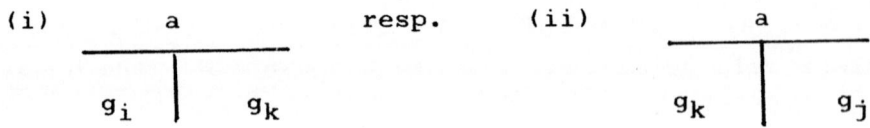

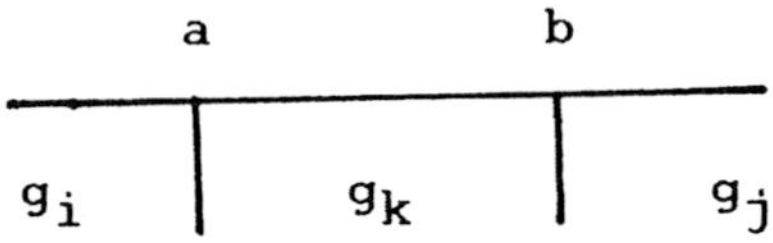

The two equations $g_k \cdot b = g_j$ and $g_j \cdot b^{-1} = g_k$ (for case (i)) resp. $g_i \cdot a = g_k$ and $g_k \cdot a^{-1} = g_i$ (for case (ii)) have to be added to the bonus list.

If finally the union of definition and bonus list contains now two equations of the form $g_i \cdot x = g_k$, $g_i \cdot x = g_{k'}$ with $k' < k$ then a <u>collapse</u> takes place:

- (α) $g_k = g_{k'}$ is added to the collapse list;
- (β) all rows with g_k in the leftmost slot are omitted;
- (γ) all occurrences of g_k in the tables, the definition and bonus list are replaced by $g_{k'}$.

(c) If neither (a) nor (b) is applicable the algorithm stops.

The basic fact about this algorithm is (cf. e.g. [C-M 80]):

1. Proposition:

The Todd-Coxeter algorithm terminates iff G is finite.

When the algorithm stops, then the g_i not eliminated by a collaps, $1 \leq i \leq n$, denote the n elements of G. In each table they are the entries in the first and last column; all other columns contain permutations of the g_i. This means, each g_i occurs in each column exactly once.

Furthermore for each g_i and $x \in \Sigma$ either $g_i \cdot x = g_k$ or $g_i \cdot x^{-1} = g_j$ for some j,k is a recorded equation. From this one can effectively construct the mutiplication table of G.

For the comparison with the completion algorithm it is convenient to introduce a notational modification:
The symbols g_n are replaced by words. The initial symbol g_1 becomes the empty word e; the words introduced in steps (2b)(i) resp. (2b)(ii) are

	a
u	ua

	a
ua^{-1}	u

The group elements are represented by words $u_1, \ldots, u_n$; because the mutiplication between words is concatenation the introduced equations are equations between words.
All equations from the definition list are trivial identities (they are introduced this way and the applications of collapse steps does not change this). Now we transfer the equations into reductions:

2. Definition:

The Todd - Coxeter system R_{TC} contains the reductions $ux \to v$ where $x \in \Sigma \cup \Sigma^{-1}$ and $u \cdot x = v$ is from the bonus list.

Consider now the example $\Sigma = \{A\}$, $A^3 = 1$:

	A	A	A
1	A	A^2	1
A	A^2	1	A
A^2	1	A	A^2

The only nontrivial equation is the bonus equation $A^3 = 1$. Both ways of reading this as a reduction ($A^3 \to 1$ resp. $1 \to A^3$) would leave each word in the inverses of A irreducible. For this reason we introduce the following extension into the algorithm:

After each bonus equation we insert the corresponding definition equation (in this case $1 \cdot A^{-1} = A^2$); the word on the right side of this equation has also to appear in the first slot of some row.

Given any word w such that $w \equiv u_i$ in G the application of R_{TC} to w from left to right reduces as above w to u_i. It is however not the case that R_{TC} is complete. In fact, it may even fail to have finite termination:

Suppose a word ubvcw is introduced as some g_k before b and c are introduced; assume furthermore that later $b = ubvcw$

and c = ubvcw and no other equation with ubvcw appear in the collapse list. Then the collapse eliminates both b and c but in the final tables we have entries

	b
1	ubvcw

and

	c
1	ubvcw

which says that b → ubvcw, c → ubvcw ∈ R_{TC} i.e. R_{TC} does not have FTP.

Restoring FTP by reversing the reductions would on the other hand violate CR.
Responsible for this effect is the rule which omits in case of a collapse those words which are introduced last. We will now modify the algorithm by allowing strategies which select the word to be omitted.

Suppose $\prec$ is any well-founded term ordering on the words.

<u>3. Definition:</u>

The <u>$\prec$ - strategy</u> says:
If u ≡ v appears in the collaps list and u $\prec$ v then v is omitted resp. replaced by u.

This simplifies the handling of bonus equation somewhat because one can immediately insert the minimum of the words given by the bonus and the definition equation.

Now we have a better situation:

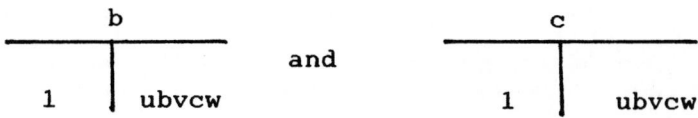

4. Proposition:

If the TC-algorithm follows the $\prec$ - strategy and generates the words $u_1,\ldots,u_n$, then,

(i) The u_i are the $\prec$ - minimal words denoting the group elements, in particular the u_i are closed under taking subwords.

(ii) R_{TC} is a complete system.

Proof: Because R_{TC} is compatible with "$\prec$" it enjoys FTP. As shown above, for any word w some application of R_{TC} (namely from left to right letter by letter) reduces w to a uniquely determined u_i. If w is irreducible this implies $w = u_i$. Assume now $u_k \; R_{TC}\xrightarrow{*} v$ we obtain $v \; R_{TC}\xrightarrow{*} u_i$ for some u_i, hence $u_k = u_i$. This shows $Irr(R_{TC}) = \{u_1,\ldots,u_n\}$.

If now in G the equation $u_k = v$, $v \prec u_k$, holds, then there is again some u_i satisfying $v \; R_{TC}\xrightarrow{*} u_i$.

This says that $u_k=u_i$ in G holds, i.e. we have i=k. Now the assertions (i) and (ii) are an immediate consequence.

[]

The system R_{TC}, although complete, may contain redundant reductions. Converting it into a minimal complete system (cf. chapter II.2.) is particularly simple because the right sides are already irreducible:

One has only to delete some reductions.

Hence there is some non-redundant complete $R_{TC}^{\,o} \subseteq R_{TC}$; this coincides with the system $R^{\infty}(\prec)$ generated by the completion algorithm. There are several differences between these methods:

a) The TC-method has a simpler termination criterion, it saves the critical pair test.

b) On the other hand the completion algorithm terminates at least not later than the TC-algorithm; in general the latter will continue if it has already some complete systems.

c) If the TC-algorithm terminates it has done two jobs: Generating a complete system and enumerating the group elements. The method of reduction systems splits this task: First a complete system is generated and then the elements are enumerated (cf. chapter IV.1.).
This has the consequence that the TC-algorithm does not terminate for any infinite group (from some point on it may generate only redundant reductions).

d) The TC-method, considered as an algorithm to generate reductions, may violate all standard principles for reduction systems like FTP or CR and still be successful.

The last remark (d) makes it desirable to compare also individual steps of both methods.

<u>(I) Discussion of a translation of [TC-steps] into [Completion steps]:</u>

We consider a defining relation $r = a_1 \dots a_n = 1$.
The bonus equation in the first row of this table just means to split the relation:

$$a_1 \dots\dots a_k = a_n^{-1} \dots\dots a_{k+1}^{-1}.$$

Rows with some word u in the first slot split the conjugate relation

$$u \cdot r \cdot u^{-1} = 1$$

They are recorded (by symmetrizing) in the completion algorithm only in the case that u is a subword of r.

In case of a collapse we have the following situation:

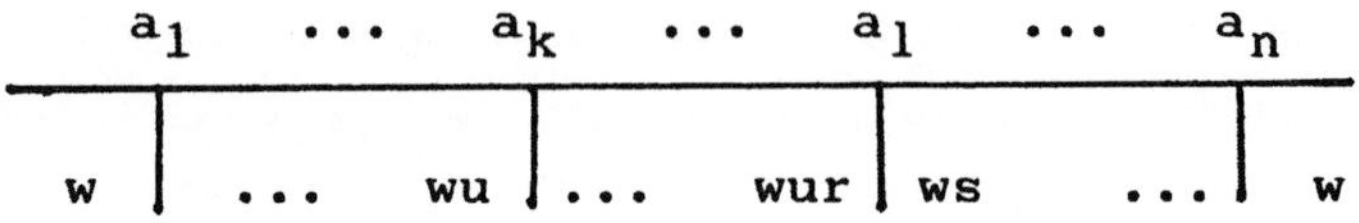

with $r = a_{k+1}\ldots a_{l-1}$. We assume the row is closed at a_l with the bonus equation $wur = wsa_l^{-1}$ and we have a collapse $wu \to v$ at a_k.
This can also be obtained either by a direct reduction or by the critical pair

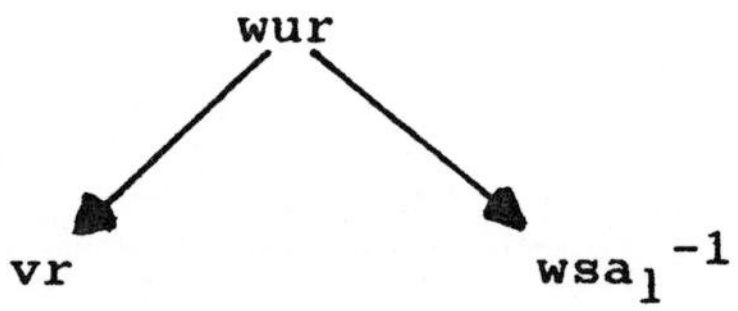

from $wu \to v$ and $ur \to sa_l^{-1}$.
The other cases are similar. This implies that all TC-steps are candidates for steps in the completion algorithm. Whether they actually apply depends again on the fact that the TC-algorithm respects the $\prec$-strategy, "$\prec$" being the order of the completion algorithm.

In addition, the completion algorithm will reduce further the critical pair.

(II) Discussion of a translation of [Completion steps] into [TC-steps]:

Because symmetrizing is the result of conjugation this can be simulated by TC-steps. Proceeding inductively one can also see that superposition can be simulated:

Assume

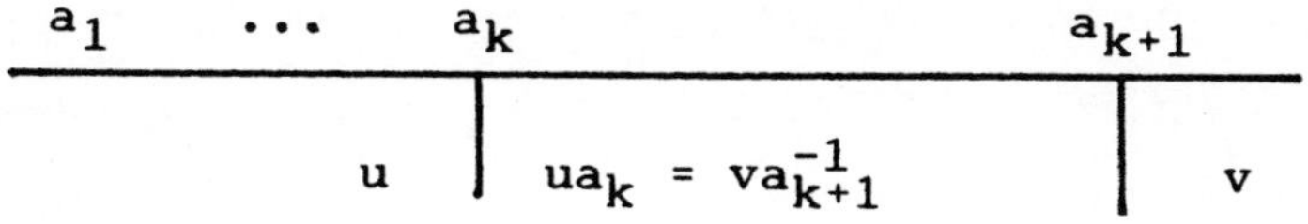

$$
\begin{array}{c|c|c}
a_1 \quad \cdots \quad a_k & & a_{k+1} \\
\hline
u & ua_k = va_{k+1}^{-1} & v
\end{array}
$$

and

$$
\begin{array}{c|c|c}
b_1 \quad \cdots \quad b_j & & b_{j+1} \\
\hline
w & wb_j = zb_{j+1}^{-1} & z
\end{array}
$$

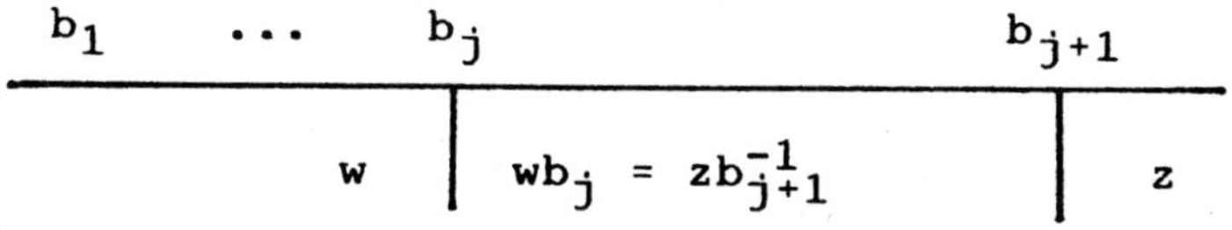

are given such that ua_k and wb_j give rise to a critical pair then similar as above we obtain this by TC-steps.

This applies, however, only to the redundant completion algorithm. Applications of reductions to left or right sides of other reductions cannot be immediately imitated. This is in particular responsible for the absence of termination in the case of an infinite group.

III. Infinite Sets of Reductions

III.1. Regular Systems

III.1.1. Regular Systems as Special Infinite Systems

In the following paragraphs we will have a closer look on nonterminating cases of the KB completion algorithms to answer questions like

- What kind of information (concerning for instance the word problem) is available if the KB completion algorithm runs forever ?

- How can we handle an infinite reduction system generated by a nonterminating KB completion algorithm ?

Nontermination of the KB completion algorithm occurs most frequently, as many experiments with an implementation of the algorithm for finitely presented groups show. Fortunately the resulting infinite system is generated very often in such a regular way that certain information is available. Consider for example a nonterminating completion algorithm generating stepwise reduction systems $\mathbf{R}_n$ for all $n \in \mathbf{N}$. If there is some computable $n_o \in \mathbf{N}$ such that for all stages $n \geq n_o$ the left sides of all reductions in $\mathbf{R}_n$ have a smaller length than the sides of all newly generated reductions (i.e. reductions from $R_{n+1}\setminus R_n$) then, of course, the word problem is solvable in the following way:

> Given words u and v one generates the reductions up to a stage where the left sides have a greater length than both u and v.

We call reduction systems with the property described as above exploding.
Remember that the completion algorithm starting with a small-cancellation-presentation of a group either terminates or explodes (see chapter II.5.).
Special cases of explosion are the regular reduction systems by which infinite systems can be handled in a finitary manner.

1. Definition:

A regular reduction scheme is a 3-tuple $(L,\sigma_{\prec},R)$ if L, R are regular expressions over the alphabet Σ with the following property

(1) $\quad l \notin \Sigma^*(L\setminus\{l,e\})\Sigma^*$ for all $l \in L$;

and if $\sigma_{\prec}: \Sigma^* \to \Sigma^*$ is a deterministic GSM-mapping with the following properties.

(2) $\quad \sigma_{\prec}$ is a bijection from L to R and undefined for all $w \in \Sigma^*\setminus L$.

(3) $\quad \sigma_{\prec}$ is compatible with the term ordering '$\prec$'. (I.e. $\sigma_{\prec}(l) \prec l$ for all $l \in L$).

If there is no need to emphasize the underlying term ordering '$\prec$' we will omit the index and write (L,σ,R).

2. Example:

Take the infinite reduction system $\{ab^n a^{-1} \to b^n \mid n \geq 1\}$ and put $L := ab^* a^{-1}$, $R := b^*$

Consider the following GSM

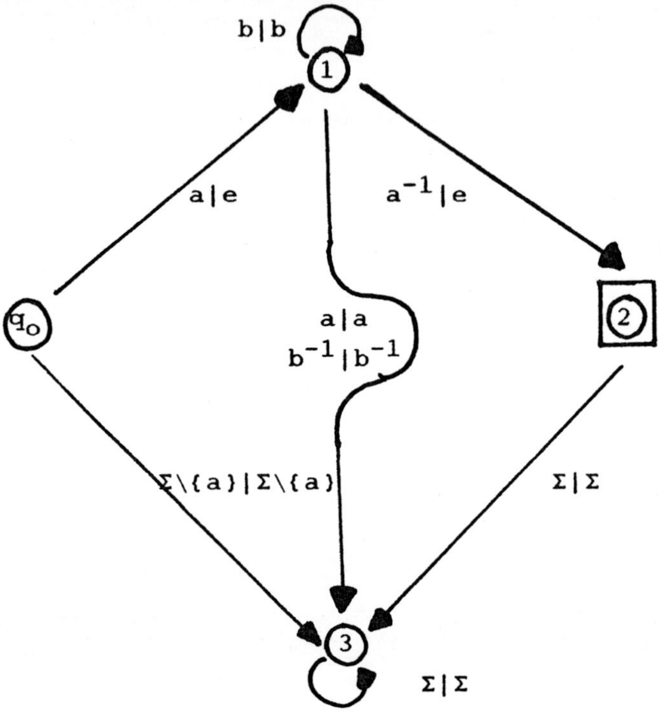

Obviously condition (1) is fulfilled for L and the conditions (2) and (3) are fulfilled for the GSM-mapping σ induced by the above GSM. Hence $(ab^* a^{-1}, \sigma, b^*)$ is a regular reduction scheme.

We have seen that a regular reduction scheme is nothing else but an eventually infinite reduction system

$$\mathbf{R} := \{l \to \sigma(l) \mid l \in L\}$$

where L as well as $R = \sigma(L)$ are regular expressions and the GSM-mapping σ (resp. σ^{-1}) preserves the regularity such that for every regular $L' \subset L$ and $R' \subset R$ the sets $\sigma(L')$ and $\sigma^{-1}(R')$ are regular too.

Moreover condition (2) states that every reduction $l \to \sigma(l) \in \mathbf{R}$ is irreducible with respect to all other reductions of **R**, whereas the condition (3) guarantees the finite termination property for **R**.

Notice also, that a usual reduction $l \to r$ can be interpreted as a reduction scheme (l, σ, r) with σ induced by the following GSM.

Let $l := l_1 l_2 \ldots l_n \in \Sigma^*$

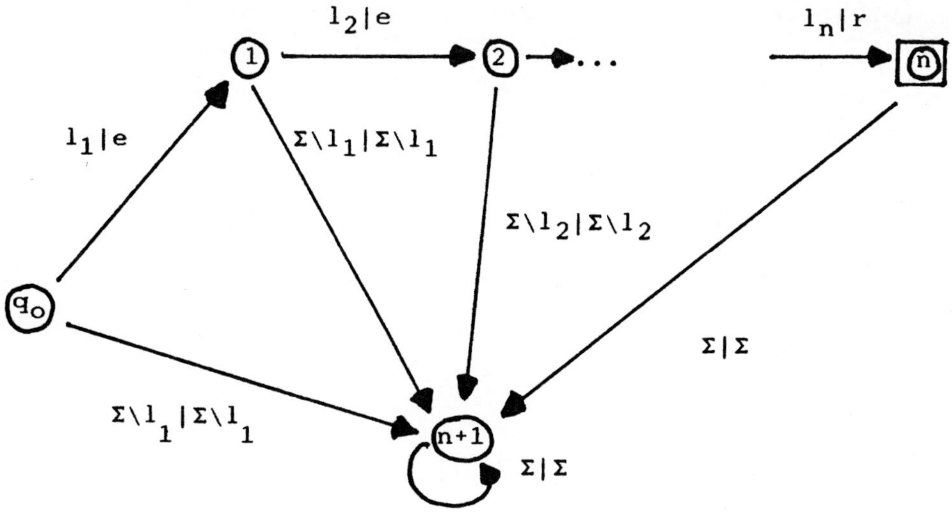

3. Definition:

We call a finite set of regular reduction schemes with the same underlying term ordering a <u>regular reduction system.</u>

Before studying regular reduction systems let us say something about their relationship to other reduction systems, the so called <u>conditional reduction systems</u>, which were investigated by Brand, Darringer and Joyner (see [BDJ 78]).

We start with the remark that a general reduction, say $x^2 \to x$, represents the infinitely many ground reductions $u^2 \to u$ for each ground word u. A more sophisticated way to denote sets of reductions is the use of conditional reductions. These are of the form

$$s \to t \quad \text{if } P$$

Here P is a predicate (with variables in s or t); the intention is that the reduction is applicable only if P is true. The language where P belongs to is quite arbitrary although it is in general some kind of first order predicate logic. The usual notions and concepts for systems of reductions can also be made here. It should be added, however, that certain aspects cannot be handled as easy as in the unconditional case (cf. [BDJ 78]). Our case of regular reduction schemes is more general but provides similar difficulties.

III.1.2. Applications of Regular Reduction Systems to Sets of Words

In order to work with regular reduction systems we obviously need an operator which reduces regular sets to regular sets by applying a regular reduction scheme.

First of all let us recall the following technical terms.

4. Definition:

Let L and T be regular expressions over Σ^*.

(i) Irr(L) := $\{t \in \Sigma^* \mid$ no $l \in L$ is a subterm of $t\}$
Irr(T,L) := Irr(L) $\cap$ T

(ii) Red(L) := $\{t \in \Sigma^* \mid$ t contains at least one $l \in L$ as a subterm$\}$
Red(T,L) := Red(L) $\cap$ T

Notice that for a regular reduction scheme (L,σ,R) the set Irr(T,L) is the set of all with respect to (L,σ,R) irreducible elements of T and Red(T,L) is the set of all with respect to (L,σ,R) reducible elements of T.
Furthermore we have in accordance with Definition 4 of chapter II.1.:

$$\text{Irr} = \text{Irr}(\mathbf{R}) = \text{Irr}(\Sigma^*,L) \quad \text{and}$$
$$\text{Red} = \text{Red}(\mathbf{R}) = \text{Red}(\Sigma^*,L) \quad \text{for } L := \{l \mid l \to r \in \mathbf{R}\}.$$

5. Lemma:

Red(T,L) and Irr(T,L) are regular whenever L and T are regular.

Proof: Obviously we have $Red(T,L) = \Sigma^* L \Sigma^* \cap T$ and
$Irr(T,L) = T \setminus Red(T,L)$.

[]

6. Definition:

A regular expression T is reducible by a regular reduction scheme (L,σ,R) to a regular expression S iff

(i) $$T = Irr(T,L) + \sum_{i=1}^{n} U_i L_i V_i$$

where U_i, L_i, V_i are regular expressions and $L_i \subseteq L$ for all $1 \le i \le n$

and

(ii) $$S = Irr(T,L) + \sum_{i=1}^{n} U_i R_i V_i$$

where $R_i = \sigma(L_i)$ for all $1 \le i \le n$.

Formally we write $T \ {}_{(L,\sigma,R)}\!\!\rightarrow S$ or $(L,\sigma,R)(T) = S$

In the following we will apply reductions only from left to right.
A formal definition of this procedure is given below.

7. Definition:

A regular expression S is the result of the regular reduction scheme (L,σ,R) applied to a regular expression T from the left iff

(i) $$T = Irr(T,L) + \sum_{i=1}^{n} U_i L_i V_i$$

where $U_i V_i L_i$ are regular, $L_i \subseteq L$ and for each $t = u_i l_i v_i \in U_i L_i V_i$ (i.e. $u_i \in U_i$, $l_i \in L_i$, $v_i \in V_i$) l_i is the first $l \in L$ which occurs in t for $1 \le i \le n$.

and

(ii) $$S = Irr(T,L) + \sum_{i=1}^{n} U_i R_i V_i$$

where $R_i = \sigma(L_i)$ for $1 \le i \le n$.

Formally we write $T \;{}_{(L,\sigma,R)}\!\rightarrow^{l} S$ or ${}_{l}(L,\sigma,R)(T) = S$ and call such a reduction also leftmost.

8. Agreement:

In the following we will consider only leftmost reductions. Hence we omit the index 'l'.

The following theorem shows that every regular reduction scheme (L,σ,R) can be applied to every regular expression.

9. Theorem:

For every regular expression T and every regular reduction scheme (L,σ,R) there is a regular expression S such that (L,σ,R)(T) = S.

Proof: Since $T = Irr(T,L) \cup Red(T,L)$ it suffices to show that $Red(T,L)$ can be written as

$$Red(T,L) = \sum_{i=1}^{n} U_i L_i V_i$$

where U_i, L_i, V_i are regular and $L_i \subseteq L$ for $1 \le i \le n$.

For that purpose we prove the following lemma.

10. Lemma:

For every regular reduction scheme (L,σ,R) there are regular expressions I_i, L_i with $L_i \subseteq Irr(L)$, $1 \le i \le n$, and

$$\sum_{i=1}^{n} L_i = L \quad \text{such that} \quad Red(L) = \sum_{i=1}^{n} I_i L_i \Sigma^*.$$

In particular we can choose the regular expressions I_i in such a way that every $t \in I_i L_i$ contains exactly one $l \in L$ as a subterm.

Proof: Of course we have $Red(L) = Irr(L)L\Sigma^*$.
A word $t \in Irr(L)L$ can however be of the form $t=t_1 l t_2$. This is the case for instance whenever there is an $l=l_1 l_2 \in L$ with

$$D_{l_1}(Irr(L)) \neq \emptyset \neq {}_{l_2}D(L)$$

i.e. $t = t_1 l_1 l_2 t_2$ with $l_1 l_2$, $l_2 t_2 \in L$ and $t_1 l_1 \in Irr(L)$.

So in order to write $Red(L)$ as the sum $\sum_{i=1}^{n} I_i L_i \Sigma^*$ with the property that for every $t = i_j l_j w_j \in I_j L_j \Sigma^*$, l_j is the first word of L which occurs in t, we must eleminate all those words from $Irr(L)L$ having not this property. Those are exactly all $ta \in Irr(L)L$ such that t contains an $l \in L$.

Put $Q := \{ta \in Irr(L)L \mid t = t_1 l t_2 \text{ for } l \in L,\ t_1,t_2 \in \Sigma^*,\ a \in \Sigma\}$.

Then we obtain according to a theorem of Brozowski:

$$Q = \sum_{a\in\Sigma} (D_a(Irr(L)L) \cap \Sigma^* L \Sigma^*)a$$

and hence Q is as the finite sum of regular sets regular.

Therefore

$$Red(L) = (Irr(L)L \backslash Q)\Sigma^*.$$

It remains to show that there are $I_i \subseteq Irr(L)$ and $L_i \subseteq L$ for $1 \le i \le n$ such that

$$Irr(L)L \backslash Q = \sum_{i=1}^{n} I_i L_i .$$

Since $Q \subseteq Irr(L)L$ we have according to Lemma 3 of chapter I.2.

$$Q = \sum_{j=1}^{k} I_j L_j$$

with $I_j := {}_{l_j}D(Irr(L)L) \subseteq Irr(L)$ and $L_j := {}_{l_j}S(Irr(L)L) \cap L \subseteq L$ for characteristic $l_1,\dots,l_k \in L$.
Because of Lemma 4 of chapter I.2. we have

$$Irr(L)L \backslash Q = Irr(L) \backslash \sum_{j=1}^{k} I_j L_j + \sum_{j=1}^{k} I_j (L \backslash L_j) =: \sum_{i=1}^{n} I_i L_i$$

where $n=k+1$ and $I_1 := Irr(L) \backslash \sum_{j=1}^{k} I_j$, $I_i := I_{i-1}$, $L_1 := L$ and $L_i := L \backslash L_{i-1}$ for $2 \le i \le k+1$.

[]

Continuation of the proof of Theorem 9:

Since $Red(T,L) = T \cap Red(L) = \sum_{i=1}^{n} (T \cap I_i L_i \Sigma^*)$, where $\sum_{i=1}^{n} I_i L_i \Sigma^*$

is the partition of Red(L) according to the last lemma, we have to show that there is also for $T_i := T \cap I_i L_i \Sigma^*$ a disjoint partition

$$T_i = \sum_{j-1}^{n-1} I_j L_j U_j$$

such that I_j, L_j, U_j are regular and $I_j \subseteq Irr(L)$, $L_j \subseteq L$ for $1 \leq j \leq n_i$ with the property that for every $t = i_j l_j u_j \in I_j L_j U_j$ (i.e. $i_j \in T_j$, $l_j \in L_j$, $u_j \in U_j$) l_j is the first word of L which occurs in t.

According to Lemma 3 of chapter I.2. we have because of $T_i \subseteq I_i L_i \Sigma^*$

$$T_i = \sum_{k=1}^{n_i} B_k U_k$$

where B_k, U_k are regular and $B_k \subseteq I_i L_i$ for $1 \leq k \leq n_i$, $1 \leq i \leq n$.
A second application of Lemma 3 of chapter I.2. to $B_k \subseteq I_i L_i$ for $1 \leq k \leq n_i$ yields the desired partition of T_i and hence of Red(T,L).

[]

11. Corollary:

Let T be regular and (L,σ,R) a regular reduction scheme and $S := \{s \in \Sigma^* \mid \exists\, t \in T$ with $t = t_1 l^{\sim} t_2$, where $l^{\sim}$ is the first word of L which occurs in t, such that $t \xrightarrow[l^{\sim} \to \sigma(l)]{} s\}$.

Then the set S is regular.

12. Definition:

Let T,S be regular expressions, (L,σ,R) a regular reduction scheme and $\mathbf{R}_{reg}$ a regular reduction system.

(a) We call T reducible to S by $\mathbf{R}_{reg}$ (formally: $S := \mathbf{R}_{reg}(T)$ or $T \; {}_{Rreg}\!\rightarrow S$) iff there is a regular reduction scheme (L,σ,R) of $\mathbf{R}_{reg}$ such that $S = (L,\sigma,R)(T)$.

(b) The n-fold application of (L,σ,R) to T (denoted by $(L,\sigma,R)^n(T)$ or $T \;{}_{(L,\sigma,R)}\!\rightarrow^n S$) is defined recursively as follows:

$$\underline{(L,\sigma,R)^n(T)} := (L,\sigma,R)((L,\sigma,R)^{n-1}(T))$$

Analogously we define:

$$\underline{\mathbf{R}^n_{reg}(T)} := \mathbf{R}_{reg}(\mathbf{R}^{n-1}_{reg}))$$

(c) $$\underline{(L,\sigma,R)^*(T)} := \bigcup_{n\in\mathbb{N}} (L,\sigma,R)^n(T)$$

is the set of all words which can be reduced from T by the regular reduction scheme (L,σ,R) and

$$\underline{\mathbf{R}^*_{reg}(T)} := \bigcup_{n\in\mathbb{N}} \mathbf{R}^n_{reg}(T)$$

is the set of all words which can be reduced from T by the regular reduction system $\mathbf{R}_{reg}$.

(d) The n-fold parallel application of (L,σ,R) to T (denoted by ${}^P(L,\sigma,R)^n(T)$) is defined recursively as follows:

$$\underline{{}^P(L,\sigma,R)^n(T)} := Irr(T,L) + \sum_{i=1}^{k} I_i R_i (L,\sigma,R)(U_i)$$

if $${}^P(L,\sigma,R)^{n-1}(T) = Irr(T,L) + \sum_{i=1}^{k} I_i R_i U_i.$$

Analogously we define:

$$\underline{P\mathbf{R}^{n}_{reg}(T)} := Irr(T,\mathbf{R}_{reg}) + \sum_{i=1}^{k} I_i R_i \mathbf{R}_{reg}(U_i)$$

if $P\mathbf{R}^{n-1}_{reg}(T) = Irr(T,\mathbf{R}_{reg}) + \sum_{i=1}^{k} I_i R_i U_i$.

(e) $\underline{P(L,\sigma,R)^{*}(T)} := \bigcup_{n\in\mathbf{N}} P(L,\sigma,R)^{n}(T_n)$

where $T_n := \{t_1 l_1 t_2 \ldots l_n t_{n+1} \in T \mid l_i \in L, t_i \in Irr(T,L)$ and l_i is the first word of L which occurs in $t_i \ldots t_{n+1}, 1 \le i \le n\}$.

is the set of words which arise when for every word $t \in T$ having n disjoint subterms $l_i \in L$, $1 \le i \le n$, these subterms l_i are replaced by $\sigma(l_i)$.

(f) $\underline{Irred(T,L)} := (L,\sigma,R)^{*}(T) \cap Irr(L)$

is the set of all with respect to (L,σ,R) irreducible words which can be deduced from T by (L,σ,R).

Analogously we define:

$\underline{Irred(T,\mathbf{R}_{reg})} := \mathbf{R}^{*}_{reg}(T) \cap Irr(\mathbf{R}_{reg})$

is the set of all with respect to $\mathbf{R}_{reg}$ irreducible words which can be deduced from T by $\mathbf{R}_{reg}$.

13. Definition:

Let T be a regular expression, (L,σ,R) a regular reduction scheme and $\mathbf{R}_{reg}$ a regular reduction system.

We call (L,σ,R) resp. $\mathbf{R}_{reg}$ <u>T-founded</u> iff there is a $n\in\mathbf{N}$ such that

$$(L,\sigma,R)^{n}(T) = (L,\sigma,R)^{n-1}(T)$$

resp. $\mathbf{R}^{n}_{reg}(T) = \mathbf{R}^{n-1}_{reg}(T)$.

14. Proposition:

For every regular expression T and every T-founded regular reduction system $\mathbf{R}_{reg}$ the sets Irred(T,L) and Irred(T,$\mathbf{R}_{reg}$) are regular.

Proof: Since (L,σ,R) and $\mathbf{R}_{reg}$ are T-founded we have for some $n \in \mathbf{N}$:

$$\text{Irred}(T,L) = (L,\sigma,R)^n(T)$$
$$\text{resp.}\quad \text{Irred}(T,\mathbf{R}_{reg}) = \mathbf{R}^n_{reg}(T)$$

This proposition follows from the fact that a deterministic GSM-mapping maps regular sets to regular sets.

If the regular reduction scheme (L,σ,R) resp. the regular reduction system $\mathbf{R}_{reg}$ is not T-founded, the sets Irred(T,L) resp. Irred(T,$\mathbf{R}_{reg}$) are generally no longer regular provided T was regular.

[]

15. Example: Let $T := (ab)^*$ and $(L,\sigma,R) := ba \to ab$.

Then one has
$$\text{Irr}((ab)^*,ba) = e + ab$$
$$\text{Red}((ab)^*,ba) = abab(ab)^*$$

Thus

$$(ba \to ab)((ab)^*) = e + ab + a^2b^2(ab)^*$$
$$(ba \to ab^2)((ab)^*) = e + ab + a^2b^2 + a^2bab^2(ab)^*$$

.

.

.

$$(ba \to ab)^n((ab)^*) = e + ab + a^2b^2 + \ldots + a^{k+1}b^{k+1} + a^{k+1}b^{k+1-m_k}ab^{m_k+1}(ab)^*$$

for $n=(k(k+1))/2+m_k$ and $0 \le m_k \le k+1$.

Hence we have for $1 \le m_k \le n+1$:

$$(ba \to ab)^*((ab)^*) = e + \sum_{n=1}^{\infty} a^n b^n + \sum_{n=2}^{\infty} a^n b^{n-m_k} ab^{m_k+1}(ab)^*$$

Therefore $\mathrm{Irred}((ab)^*, ba \to ab) = \{a^n b^n \mid n \in \mathbb{N}\}$ is context-free but no longer regular.

<u>16. Example:</u> Let $T := (abc)^*$ and (L,σ,R) be defined as follows:

$$L := ba + ca + cb, \qquad R := ab + ac + bc$$

and $\sigma(ba) := ab$, $\sigma(ca) := ac$, $\sigma(cb) := bc$.

Analogously to the last example we have

$$\mathrm{Irred}(T,L) = \{a^n b^n c^n \mid n \in \mathbb{N}\}$$

which is not even context-free.

Nevertheless we gain in many cases regular expressions for Irred(T,L) and $\mathrm{Irred}(T,\mathbf{R}_{reg})$ even when the regular reduction scheme (L,σ,R) resp. the regular reductions scheme $\mathbf{R}_{reg}$ is not T-founded, provided of course T is regular.

In order to characterize these cases it is useful to know the following lemma. It says that the set of words which arises by reducing parallelly every word of a regular expression is again regular. (Remember that producing a word in parallel means that, while running once from left to right through t, t is reduced as often as possible).

17. Lemma:

Let T be a regular expression (L,σ,R) a regular reduction scheme and $\mathbf{R}_{reg}$ a regular reduction system. Then $P(L,\sigma,R)^*(T)$ and $P\mathbf{R}_{reg}{}^*(T)$ are regular.

Proof: In order to show that $P(L,\sigma,R)^*(T)$ is regular we construct a GSM **σ**, whose GSM-mapping **σ** has the following property

$$\boldsymbol{\sigma}(T) = P(L,\sigma,R)^*(T)$$

for all regular T. This implies the first proposition.

The second proposition can be proved analogously.

Let $M_{Irr} = (Q_{Irr}, \varphi_{Irr}, q_o, F_{Irr})$ be a finite deterministic automaton accepting Irr(L) and $\sigma = (Q,\Sigma,\Sigma,\varphi,q_o,F)$ the deterministic GSM from (L,σ,R).

W.l.o.g. we assume $Q \cap Q_{Irr} = \{q_o\}$.

Further we define

$$Q^{\sim} := \{ q \in Q \backslash F \mid \exists w \in \Sigma^* \text{ s.t. } p \in F \text{ whenever } \varphi(q,w) = (p,v)\}$$

the set of all states which lead in one step to some halting state and put

$$\mathbf{Q} := Q^{\sim} \cup F_{Irr} \quad \text{and} \quad \mathbf{F} := F_{Irr} \cup \{q_o\}.$$

The desired GSM **σ** can be defined as follows:

$$\boldsymbol{\sigma} := (\mathbf{Q},\Sigma,\Sigma,\boldsymbol{\varphi},q_o,\mathbf{F})$$

where **φ** are the following subsets of Q x Σ x Q x Σ

(i) $\boldsymbol{\varphi}(q_o,a) := \{(\varphi_{Irr}(q_o,a),a),\ \varphi(q_o,a)\}$
for all $a\in\Sigma$ s.t. $\varphi_{Irr}(q_o,a)\in F_{Irr}$;

(ii) $\boldsymbol{\varphi}(q,a) := \{(\varphi_{Irr}(q,a),a),\ \varphi(q_o,a)\}$
if $q\in F_{Irr}\backslash\{q_o\}$ and $\varphi_{Irr}(q,a)\in F_{Irr}$, $a\in\Sigma$;

(iii) $\boldsymbol{\varphi}(q,a) := \varphi(q,a)$
if $q\in Q^{\sim}\backslash\{q_o\}$ and $p\notin F$ for $\varphi(q,a)=(p,v)$;

(iv) $\boldsymbol{\varphi}(q,a) := (s,v)$
if $q\in Q^{\sim}\backslash\{s\}$ and $p\in F$ for $\varphi(q,a)=(p,v)$.

[]

<u>18. Theorem:</u>

Let (L,σ,R) be a regular reduction scheme.
Then Irred(T,L) is regular for every regular expression T whenever $\beta(L)\cap\varepsilon(R) = \emptyset = \varepsilon(L)\cap\beta(R)$.

(Remember: $\beta(T)$ resp. $\varepsilon(T)$ is the set of initial parts resp. end parts of T (see Definition 5 of chapter I.2.))

<u>Proof:</u> Let $(L,\sigma,R)(T) = \mathrm{Irr}(T,L) + \sum_{i=1}^{n} I_iR_iU_i$

with $I_i \subseteq \mathrm{Irr}(L)$ and $R_i=\sigma(L_i)$ for $1\le i\le n$.
The condition $\beta(L)\cap\varepsilon(R) = \emptyset = \varepsilon(L)\cap\beta(R)$ implies

$$\mathrm{Irr}(T,L) + \sum_{i=1}^{n} I_iR_i \subseteq \mathrm{Irr}(L)$$

Hence it can easily be shown by induction that

$$\mathrm{Irred}(T,L) = P(L,\sigma,R)^{*}(T).$$

Now the proposition follows obviously by the last lemma. []

19. Theorem:

Let T be a regular expression and (L,σ,R) a regular reduction scheme.
Then Irred(T,L) is regular whenever $(L,\sigma,R)(T) \subseteq T$.

Proof: $(L,\sigma,R)(T) \subseteq T$ implies $(L,\sigma,R)^n(T) \subseteq T$ for all $n \in \mathbb{N}$. Hence we have $(L,\sigma,R)^*(T) \subseteq T$, what implies:

$$\mathrm{Irred}(T,L) = (L,\sigma,R)^*(T) \cap \mathrm{Irr}(L) \subseteq T \cap \mathrm{Irr}(L) = \mathrm{Irr}(T,L)$$

Together with $\mathrm{Irred}(T,L) \supseteq \mathrm{Irr}(T,L)$ we obtain

$$\mathrm{Irred}(T,L) = \mathrm{Irr}(T,L),$$

which of course is regular. []

Conclusion:

For each free monoid T and each regular reduction scheme (L,σ,R) with $R \subseteq T$ Irred(T,L) is regular.

20. Theorem:

Let T be a regular expression and (L,σ,R) a regular reduction scheme:

(i) If $(L,\sigma,R)(T) = I+TS$ with $IS^* \subseteq \mathrm{Irr}(L)$ and $I=\mathrm{Irr}(T,L)$ then $\mathrm{Irred}(T,L) = IS^*$ (This is of course regular)

(ii) If $(L,\sigma,R)(T) = I+ST$ with $S^*T \subseteq \mathrm{Irr}(L)$ and $I=\mathrm{Irr}(T,L)$ then $\mathrm{Irred}(T,L) = S^*I$
(This is again of course regular)

Proof: (i) Since $(L,\sigma,R)(T)=I+TS$ there is for every $n\in\mathbf{N}$ a $t\in T$ such that $(L,\sigma,R)^n(t)\neq(L,\sigma,R)^{n-1}(t)$, i.e. $(L,\sigma,R)^{n-1}(t)$ is not irreducible with respect to (L,σ,R).

Let

$$T_n := \{t\in T \mid (L,\sigma,R)^n(t) = (L,\sigma,R)^{n+1}(t) \text{ and } (L,\sigma,R)^n(t) \neq (L,\sigma,R)^{n-1}(t)\}.$$

Then we have:

$$T = \sum_{n=0}^{\infty} T_n \qquad \text{and} \qquad (L,\sigma,R)^*(T) = \sum_{n=0}^{\infty} (L,\sigma,R)^n(T_n).$$

1. Statement: $(L,\sigma,R)^n(T) = \sum_{i=0}^{n-1} IS^i + TS^n$ for all $n\in\mathbf{N}$.

Proof by induction on n:

Assume Statement 1 is true for n=k. Then we have

$$\begin{aligned}(L,\sigma,R)^{k+1}(T) &= (L,\sigma,R)((L,\sigma,R)^k(T))\\ &= (L,\sigma,R)\Big(\sum_{i=0}^{k-1} IS^i + TS^k\Big)\\ &= \sum_{i=0}^{k-1} IS^i + (I+TS)S^k\\ &= \sum_{i=0}^{k} IS^i + TS^{k+1}\end{aligned}$$

2.Statement: $(L,\sigma,R)^n(T_n) = IS^n$ for all $n\in\mathbf{N}$.

Proof by induction on n:

First we have

$$\begin{aligned}(L,\sigma,R)(T_1) &= ((L,\sigma,R)(T) \cap Irr(L)) \setminus I\\ &= ((I+TS) \cap Irr(L)) \setminus I\\ &= (I + Red(T,S)S) \cap Irr(L)\\ &= IS\end{aligned}$$

Assume statement 2 to be true for n=k. Then we have:

$$(L,\sigma,R)^{k+1}(T_{k+1}) = ((L,\sigma,R)^{k+1}(T)\cap Irr(L)) \setminus \sum_{i=1}^{k} (L,\sigma,R)^{i}(T_i)$$

$$= (\sum_{i=0}^{k} IS^{i} + TS^{k+1}) \cap Irr(L) \setminus \sum_{i=0}^{k} IS^{i}$$

$$= TS^{k+1} \cap Irr(L) = IS^{k+1}.$$

Hence we obtain:

$$(L,\sigma,R)^{*}(T) = \sum_{n=0}^{\infty} (L,\sigma,R)^{n}(T_n) = \sum_{n=0}^{\infty} IS^{n} = IS^{*}$$

(ii) analogously to (i)

[]

<u>21. Theorem:</u>

Let (L,σ,R) be a regular reduction scheme, T a regular expression and T_n defined as in the proof of Theorem 20. Then we have:

If $(L,\sigma,R)^{n}(T_n) = \sum_{i=0}^{k} U_i S_i^n V_i$, for all $n>m\in\mathbb{N}$ and regular U_i, S_i, V_i, $1\leq i\leq k$,

then Irr(T,L) is regular.

<u>Proof:</u>

$$Irr(T,L) = \sum_{n=0}^{\infty} (L,\sigma,R)(T_n)$$

$$= \sum_{i=0}^{m-1} (L,\sigma,R)^{i}(T_i) + \sum_{j=m}^{\infty} (L,\sigma,R)^{j}(T_j)$$

$$= \sum_{i=0}^{m-1} (L,\sigma,R)^i(T_i) + \sum_{j=m}^{\infty} \sum_{i=0}^{k} U_i S_i^j V_i$$

$$= \sum_{i=0}^{m-1} (L,\sigma,R)^i(T_i) + \sum_{i=0}^{k} \sum_{j=m}^{\infty} U_i S_i^j V_i$$

$$= \sum_{i=0}^{m-1} (L,\sigma,R)^i(T_i) + \sum_{i=0}^{k} \sum_{j=0}^{\infty} U_i S_i^m S_i^j V_i$$

$$= \sum_{i=0}^{\infty} (L,\sigma,R)^i(T_i) + \sum_{i=0}^{\infty} U_i S_i^m S_i^* V_i$$

Since $(L,\sigma,R)^i(T_i)$ is regular for all $i\in\mathbb{N}$ and in particular all sets $U_i S_i^m S_i^* V_i$ are regular for $0\leq i\leq k$, everything is proved.

[]

III.1.3. The Undecidability of the Church-Rosser Property

Working with regular reduction systems the Church-Rosser property is of course of great interest.
Unfortunately we have the following theorem:

22. Theorem:

It is undecidable whether an infinite regular reduction system is complete.

Proof: The following proof is borrowed from O'Dunlaing ([O'Du 83]), who proved the undecidability of the completeness of infinite reduction systems $\mathbf{R}_D$ of the form:

$\mathbf{R}_D := \{w_{ij} \to a_i \mid a_i \in \Sigma,\ w_{ij} \in R_i,\ R_i \text{ is regular},\ 1 \leq i \leq m,\ j \in \mathbf{N}\}$.

As O'Dunlaing we proceed from the fact that for every Turing machine TM there is a finite reduction system $\mathbf{R}_{TM}$ corresponding to TM (see M. Davis [Da 58], pages 88 - 93).

Proceeding from $\mathbf{R}_{TM}$ we can construct a well-founded regular reduction system $\mathbf{R}(w)$ (w any word of Σ^*), which is complete iff TM stops on input w.
The unsolvability of the halting problem for TM yields immediately the undecidability of the Church-Rosser property of $\mathbf{R}(w)$.

In particular, one can (eventually by adding some dummy

symbols) construct a finite reduction system R_{TM} corresponding to the Turing machine TM in such a way that the following properties are fulfilled:

Let Σ be the input alphabet of TM

(a) There is a regular set CONF corresponding to the set of configurations of TM and a map f: CONF $\rightarrow$ set of configurations of TM such that

$$t \; R_{TM}\rightarrow^{*} \; s \quad \text{iff} \quad f(s) \; {}_{TM}\vdash^{*} \; f(t)$$

(b) Any $t \in$ CONF can overlap some $l \in Left(R_{TM})$ only if l is a subterm of t.
Any $t \in$ CONF can overlap some $s \in$ CONF if and only if t and s are identical.

(c) R_{TM} is complete.

(d) For every $w \in \Sigma^{*}$ there is a with respect to R_{TM} irreducible word $w^{*} \in$ CONF, such that the Turing machine TM stops on input w iff there is some $h \in$ HALT (HALT := f^{-1}(set of halting configurations of TM)) with $h \; R_{TM}\rightarrow^{*} \; w^{*}$.

Let $\Omega \supseteq \Sigma$ be the alphabet of R_{TM} and Ψ an alphabet which is disjunct but isomorph to Ω where $g: \Omega \rightarrow \Psi$ is a bijection. Then we define a regular reduction system R(w) for some $w \in \Sigma^{*}$, $w \notin$ HALT as following:

$$R(w) := R_{TM} \cup R_{TM}^{\sim} \cup \{l \rightarrow g(l) \mid l \in M_w\},$$

$$\text{where } M_w := \text{HALT} \cup \{t \in \text{KONF} \mid t \neq w^{*},\ t \text{ irreducible}\}$$
$$\text{and } R_{TM}^{\sim} := \{g(l) \rightarrow g(r) \mid l \rightarrow r \in R_{TM}\}$$

<u>Proposition:</u> **R**(w) is incomplete iff the Turing mashine TM stops on input w.

<u>Proof:</u> Because of the properties (a)-(c) **R**(w) is incomplete iff there is some $h \in M_w$ such that $\mathrm{Irred}(h,\mathbf{R}_{TM}) \notin M_w$. But this is only true iff $\mathrm{Irred}(h,\mathbf{R}_{TM}) = w^*$, i.e. iff the Turing machine TM stops on input w.

[]

Additional remarks:

In [O'Du 83] a different kind of regular systems has been studied where the left sides form a regular set and the right sides consist of only one single-letter word. It was shown that CR was undecidable for such systems. The kind of regular systems studied in chapter III.1 naturally arise as limit systems of the completion algorithm. For instance, for the KB-orderings the van Dyck groups D(n,m,k) defined by $A^k = B^m = (AB)^n = 1$ have a complete regular system. The infinite cases are exactly the following:

1) $k=2$, $m=5$ and $n=2j$ for some $j>1$.
2) $k\geq5$, $m=2$ and $n=2j$ for some $j>1$.
3) $k,m \in \{2,3\}$, $k\neq m$ and $n\geq6$.
4) $k,m \in \{3,4\}$, $k\neq m$ and $n=2j$ for some $j\geq2$.
5) $k=2$, $m=4$ and $n\geq4$.
6) $k=3$, $m=3$ and $n\geq3$.
7) $k=4$, $m=2$ and $n=2j$ for some $j\geq2$.

Groups with undecidable word problem cannot have a complete system, however. In chapter IV.5 we will see an example of a group for which $\mathbf{R}^\infty$ is not regular for any KB-ordering.

III.1.4. A Possible Church-Rosser Test

Although the Church-Rosser property is undecidable for infinite regular reduction systems $\mathbf{R}_{reg}$, even if they possess the finite termination property, it is possible to postulate some criteria which are sufficient for the Church-Rosser property of $\mathbf{R}_{reg}$.
In the following we will show that under certain conditions the usual test for all $w \in \text{Superpos}(\text{Left}(\mathbf{R}_{reg}),\text{Left}(\mathbf{R}_{reg}))$ will decide whether $\mathbf{R}_{reg}$ is complete or not.

23. Definition:

For regular expression L_1, L_2 we define:
$\underline{\text{Superpos}(L_1,L_2)} := \{w \in \Sigma^* \mid \exists\ u \in \Sigma^*,\ l_1 u \in L_1,\ u l_2 \in L_2$ such that $w = l_1 u l_2\}$.

24. Theorem:

(i) Superpos (L_1,L_2) is regular for regular expressions L_1, L_2.

(ii) For two regular reduction schemes (L_1,σ_1,R_1) and (L_2,σ_2,R_2) Superpos(L_1,L_2) splits in particular into the following sums:

$$\text{Superpos}(L_1,L_2) = \sum_{i=1}^{n} L_{1i}E_i = \sum_{j=1}^{k} B_j L_{2j}$$

such that $L_{1i} \subseteq L_1$, $E_i \subseteq \varepsilon(L_2)$, $B_j \subseteq \beta(L_1)$ and $L_{2j} \subseteq L_2$ for $1 \le i \le n$, $1 \le j \le k$.

<u>Proof:</u> (i) Obviously we have:

$$\mathrm{Superpos}(L_1,L_2) = L_1\varepsilon(l_2) \cap \beta(L_1)L_2.$$

Hence proposition (i) follows immediately from the regularity of initial parts resp. end parts.

(ii) Since $\mathrm{Superpos}(L_1,L_2) \subseteq L_1\varepsilon(L_2)$, Theorem 3 of chapter I.2. yields:

$$Q := \mathrm{Superpos}(L_1,L_2) = \sum_{i=1}^{n} l_{1i}E(Q)_{l1i}D(Q) = \sum_{i=1}^{n} L_{i1}E_{2i}$$

The second part of the proposition can be proved analogously.

[]

<u>25. Definition:</u>

Let (L_1,σ_1,R_1) and (L_2,σ_2,R_2) be regular reduction schemes and

$$\mathrm{Superpos}(L_1,L_2) = \sum_{i=1}^{n} L_{1i}E_i = \sum_{j=1}^{k} B_jL_{2j}$$

be given as in Theorem 22.

We call (T_1,T_2) a <u>critical pair</u> for the reduction schemes (L_1,σ_1,R_1) and (L_2,σ_2,R_2) if

$$T_1 := \sum_{i=1}^{n} \sigma_1(L_{1i})E_i$$

and
$$T_2 := \sum_{j=1}^{k} B_j\sigma_2(L_{2j}).$$

26. Definition:

A regular expression T is called separated with respect to **R** if $Irred(v,\mathbf{R}) \cap Irred(w,\mathbf{R}) = \emptyset$ for all $v,w \in T$, $v \neq w$.

Notice that in particular every set T with $|T| = 1$ is, of course, separated.

As the following theorem shows there is a finite test deciding the Church-Rosser property for a regular reduction system

$$\mathbf{R}_{reg} = \{(L_i,\sigma_i,R_i) \mid 1 \le i \le n\}$$

with the property that every $Superpos(L_i,L_j)$, $1 \le i,j \le n$ can be represented as a finite sum of regular sets which are separated with respect to $\mathbf{R}_{reg}$.

27. Theorem:

Let $\mathbf{R}_{reg} = \{(L_i,\sigma_i,R_i) \mid 1 \le i \le n\}$ be a regular reduction system.
If for all regular separated sets $T \subseteq Superpos(L_i,L_j)$, $1 \le i,j \le n$

$$Irred(T_1,\mathbf{R}_{reg}) = Irred(T_2,\mathbf{R}_{reg})$$

where (T_1,T_2) is the critical pair belonging to T, then we have

$$|Irred(w,\mathbf{R}_{reg})| = 1 \quad \text{for all } w \in \Sigma^*.$$

Proof: Assume there is some $w \in \Sigma^*$ such that

$$|Irred(w,\mathbf{R}_{reg})| > 1.$$

W.l.o.g. we consider w to be minimal according to the underlying term ordering $\prec$.

Then we have $w = l_1ul_2$, i.e. there are (L_1,σ_1,R_1), $(L_2,\sigma_2,R_2) \in \mathbf{R}_{reg}$ such that $l_1u \in L_1$ and $ul_2 \in L_2$.

Let T be a regular separated subset of $Superpos(L_1,L_2)$ with $l_1ul_2 \in T$ and let (T_1,T_2) be the critical pair belonging to T.

Furthermore let $w_1 \in Irred(r_1l_2,\mathbf{R}_{reg})$ and $w_2 \in Irred(l_1r_2,\mathbf{R}_{reg})$, where $r_1 = \sigma_1(l_1u)$ and $r_2 = \sigma_2(ul_2)$,
i.e. we have $w_1 \in Irred(T_1,\mathbf{R}_{reg})$, $w_2 \in Irred(T_2,\mathbf{R}_{reg})$.

Since $Irred(T_1,\mathbf{R}_{reg}) = Irred(T_2,\mathbf{R}_{reg})$ we have $w_1 \in Irred(T_2,\mathbf{R}_{reg})$ which implies that there is some $l_1'u'l_2' \in T$ such that $w \in Irred(l_1'r_2',\mathbf{R}_{reg})$ with $r_2' = \sigma_2(u'l_2')$.

Hence we have $w_1 \in Irred(l_1ul_2,\mathbf{R}_{reg}) \cap Irred(l_1'u'l_2',\mathbf{R}_{reg})$.

Since T is separated we then have $l_1ul_2 = l_1'u'l_2'$ and therefore $w_1 \in Irred(l_1r_2,\mathbf{R}_{reg})$.

In the same way one shows $w_2 \in Irred(r_1l_2,\mathbf{R}_{reg})$.

The minimality of w and $w \succ l_1r_2$, $w \succ r_1l_2$ implies then

$$|Irred(l_1r_2,\mathbf{R}_{reg})| = |Irred(r_1l_2,\mathbf{R}_{reg})| = 1$$

hence w_1 is identical with w_2.

[]

If $Irred(T_1,\mathbf{R}_{reg}) \neq Irred(T_2,\mathbf{R}_{reg})$ for the critical pair (T_1,T_2) belonging to some $T = Superpos(L_1,L_2)$ then, obviously, $\mathbf{R}_{reg}$ is incomplete. The reversal however is not true, as the following example shows.

<u>Example:</u> Let $\mathbf{R}_{reg} := \{(L_i, \sigma_i, R_i) \mid 1 \le i \le 4\}$ be given as follows:

$L_1 := ca + bd,$ $\qquad R_1 := x + y,$

$L_2 := ab + dc,$ $\qquad R_2 := y + x,$

$L_3 := x(a + b + c + d),$ $\qquad R_3 := (a + b + c + d)x,$

$L_4 := y(a + b + c + d + x),$ $\qquad R_4 := (a + b + c + d + x)y,$

$\sigma_1(ca) = x, \quad \sigma_1(bd) = y$

$\sigma_2(ab) = y, \quad \sigma_2(dc) = x$

$\sigma_3(xa) = ax, \quad \sigma_3(xb) = bx, \quad \sigma_3(xc) = cx, \quad \sigma_3(xd) = dx$

$\sigma_4(ya) = ay, \quad \sigma_4(yb) = by, \quad \sigma_4(yc) = cy, \quad \sigma_4(yd) = dy,$

$\sigma_4(yx) = xy$

Then we have for instance

$$\text{Superpos}(L_1, L_2) = cab + bdc =: T$$

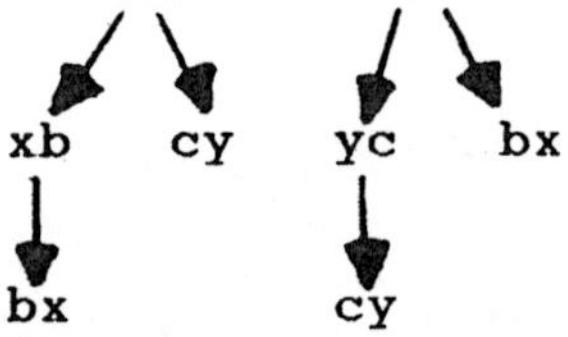

and $\text{Irred}(T_1, \mathbf{R}_{reg}) = \text{Irred}(T_2, \mathbf{R}_{reg})$, where (T_1, T_2) with $T_1 = xb + yc$ and $T_2 = cy + bx$ is the critical pair belonging to T.

Testing all other $T := \text{Superpos}(L_i, L_j)$ yields also

$$\text{Irred}(T_1, \mathbf{R}_{reg}) = \text{Irred}(T_2, \mathbf{R}_{reg})$$

for the critical pairs (T_1, T_2) belonging to T.

Nevertheless there is some $w \in \Sigma^*$, for instance, $w = cab$, such that $|\text{Irred}(w, \mathbf{R}_{reg})| > 1$.

There are however incomplete regular reduction systems with the property that for all $T := \text{Superpos}(L_i, L_j)$ we have $\text{Irred}(T_1, \mathbf{R}_{reg}) = \text{Irred}(T_2, \mathbf{R}_{reg})$ for the critical pair (T_1, T_2) belonging to T.
Hence the condition of being separated is in fact necessary for the correctness of the KB-completion-test.

The assumptions under which the completeness of a regular reduction system is decidable by the KB-completion-test are listed in the following corollary.

28. Corollary:

Let $\mathbf{R}_{reg} := \{(L_i, \sigma_i, R_i) \mid 1 \le i \le n\}$ be a regular reduction system.
If very $T = \text{Superpos}(L_i, L_j)$, $1 \le i, j \le n$ can be decomposed into the sum of finitely many with respect to $\mathbf{R}_{reg}$ separated sets

$$T = \sum_{k=1}^{m} T_k$$

such that for every critical pair (T_{1k}, T_{2k}) belonging to T_k one can decide whether

$$\text{Irred}(T_{1k}, \mathbf{R}_{reg}) = \text{Irred}(T_{2k}, \mathbf{R}_{reg})$$

then it is decidable whether $\mathbf{R}_{reg}$ is complete.

Notice that generally it is not decidable whether a set T is separated and that $\text{Irred}(T, \mathbf{R}_{reg})$ can not always be computed. Nevertheless there are many regular reduction systems which satisfy the assumptions of Corollary 28. Sufficient conditions for the regularity and in particular the computability of $\text{Irred}(T, \mathbf{R}_{reg})$ were already given in section III.1.2.

III.2. Forward - Backward Systems

The free abelian group generated by A and B is defined by the single equation AB = BA. For the weights $\omega(A) = \omega(B) = 1$ the completion algorithm generates an infinite set of reductions:

$$BA \to AB,\ B^{-1}A^{-1} \to A^{-1}B^{-1},\ B^{-1}A \to AB^{-1},\ A^{-1}B \to BA^{-1}$$
$$\text{and}\quad B(A^{-1})^nB^{-1} \to (A^{-1})^n,\ AB^nA^{-1} \to B^n \quad \text{for} \quad n \geq 1.$$

Similarly the completion algorithm explodes for the Greendlinger group which is defined by ABC = CBA (which can be regarded as a generalized law of commutativity). From II.4. we recall:
For the weights $\omega(A) = \omega(B) = \omega(C) = 1$ the completion algorithm generates an infinite set of reductions:

$$\mathrm{SYM}(CBA \to ABC),\ A(BC)^nA^{-1} \to (CB)^n,\ B(CA^{-1})^nB^{-1} \to (A^{-1}C)^n$$
$$C(A^{-1}B^{-1})^nC^{-1} \to (B^{-1}A^{-1})^n,\ B(AC^{-1})^nB^{-1} \to (C^{-1}A)^n \text{ for } n\geq 1$$

(We have also seen that there is no finite complete system in any KB-ordering for this group.)

In both examples the complete sets are regular. Although the regular expressions suffice to describe the infinite systems they do not reflect the fact that they contain many reductions which become redundant after some trivial manipulation. If e.g. one would be able to apply only a few reductions from right to left both infinite systems would collapse to finite ones. This suggests the idea to reintroduce symmetry in some restricted and controlled manner.

This has to be done in such a way that the concepts of a reduction system does not loose those features which makes it a powerful tool for decision problems.
The notion of a forward-backward system ((f+b)-system in short) captures partly this idea. A finite (f+b)-system can sometimes simulate an infinite ordinary reduction system.
The general concepts will be defined for arbitary term algebras.

1. Definition:

An <u>(f+b)-system</u> (R_f+R_b) is an ordered pair (R_f,R_b) of ordinary reduction systems compatible with well-founded term orderings $\prec_f$ and $\prec_b$ resp..

R_f is called the <u>forward system</u> and R_b is the <u>backward system</u>; both are in general assumed to be finite. Instead of $_{Rf}\rightarrow$ and $_{Rb}\rightarrow$ we write $_f\rightarrow$ and $_b\rightarrow$ in short.

2. Definition:

(i) A <u>repetition free derivation</u> of a term v from a term u in the system $R = R_f \cup R_b$ is a sequence $\langle u_i \mid 1 \le i \le n \rangle$ such that $u_i \ _R\rightarrow u_{i+1}$ for $1 \le i \le n$, $u_1 = u$, $u_n = v$ and $u_i \neq u_k$ for $i \neq k$.

(ii) An <u>(f+b)-derivation</u> of v from u is a repitition free derivation of v from u such that $v \prec_f u$.
Notation: $u \ _{f+b}\overset{*}{\rightarrow} v$

As usual $u \ _{f+b}\overset{*}{\rightarrow} v$ means the existence of some (f+b)-derivation from u to v. Because repititions are always redundant

this is equivalent to $u \ {}_{R}\!\!\to^{*} v$ and $v \prec_f u$.
We refer to (f+b)-derivations as to derivations in (R_f+R_b).
Note also that these concepts are not symmetric.

Having defined $u \ {}_{f+b}\!\!\to^{*} v$, we can formally take over the notions of irreducible and reducible terms from chapter II.1..
Furthermore, the concept of the Church-Rosser property CR will be used in the same way as in chapter II.1., i.e. the square

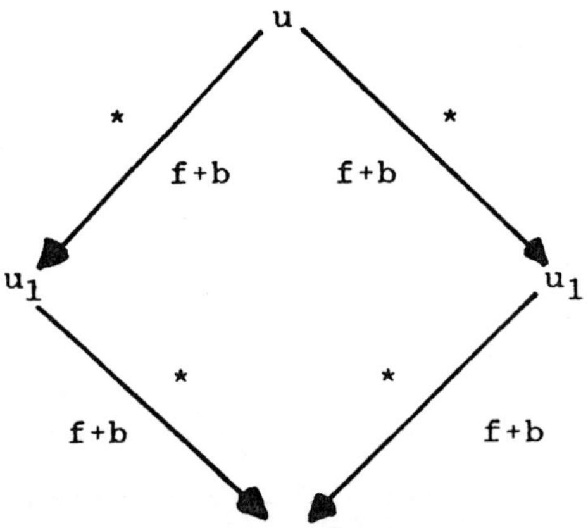

can be completed for given u, u_1 and u_2.

It is however not useful to take over the formal equivalents of the properties WCR and FTP. This is due to the fact that (f+b)-derivations are not arbitrary iterations of elementary reduction steps but are subject to the condition of being repitition free. The substitution for WCR will be discussed later; we first focus on FTP.

3. Definition:

(R_f+R_b) has weak FTP iff there are no infinite repitition free derivations in $R = R_f \cup R_b$.

If (R_f+R_b) enjoys weak FTP then by König's lemma for each term u the sets $L(u,R_f+R_b) := \{v \mid u \xrightarrow[f+b]{*} v\} \cup \{u\}$ and $\{v \mid v$ can be reached from u by a repitition free derivation$\}$ are both finite.
An immediate consequence is

4. Proposition:

Suppose (R_f+R_b) has weak FTP. Then

(i) Irreducibility is a decidable property of terms.

(ii) For each term the sets $L(u,R_f+R_b) \cap Irr = Irred(u,R_f+R_b)$ and $L(u,R_f+R_b)$ are finite and computable from u.

Completeness is defined as CR + weak FTP, UTP again means that $L(u,R_f+R_b)$ contains exactly one term. As for ordinary systems weak FTP guarantees the equivalence of CR and UTP. An immediate consequence is that a complete (f+b)-system solves the word problem.

Our next examples will show finite (f+b)-systems which are derivation equivalent to the infinite systems given at the beginning of this section.

1) The free abelian group on two generators A and B:

We put $R_f = \{BA \rightarrow AB,\ B^{-1}A^{-1} \rightarrow A^{-1}B^{-1},\ B^{-1}A \rightarrow AB^{-1},\ A^{-1}B \rightarrow BA^{-1}\}$ and define R_b to contain the reversed reductions.

The orderings are the KB-orderings with weights $\omega(A)=\omega(B)=1$ in both cases; we choose the lexicographic ordering with A preceding B for $<_f$; in $<_b$ B precedes A.

2) <u>The Greendlinger group:</u>

We put $\mathbf{R}_f = \{ \quad CBA \to ABC, \quad A^{-1}CB \to BCA^{-1},$
$C^{-1}B^{-1}A^{-1} \to A^{-1}B^{-1}C^{-1}, \quad B^{-1}C^{-1}A \to AC^{-1}B^{-1},$
$C^{-1}AB \to BAC^{-1}, \quad B^{-1}A^{-1}C \to CA^{-1}B^{-1} \};$

$\mathbf{R}_b$ again contains the reversed reductions.
The orderings are taken as in the above example with $A \prec_f B \prec_f C$ and $C \prec_b B \prec_b A$.

Both (f+b)-systems have CR and are derivation equivalent to infinite ordinary systems from above.
They also enjoy weak FTP because right and left sides have the same weights and for each weight there are only finitely many words having that weight.
Because for any KB-ordering there is no finite complete system we see that (f+b)-systems are more powerful than ordinary systems.

The formation of $(\mathbf{R}_f+\mathbf{R}_b)$ from two systems $\mathbf{R}_f$ and $\mathbf{R}_b$ which both have FTP may loose even weak FTP, however. As an example we consider

$\mathbf{R}_f = \{A \to BB\}$ with weights $\omega_f(A) = 3, \omega_f(B) = 1;$
$\mathbf{R}_b = \{B \to AA\}$ with weights $\omega_b(A) = 1, \omega_b(B) = 3.$

In $(\mathbf{R}_f+\mathbf{R}_b)$ we have the infinite repetition free derivation

$$A \to BB \to AAAA \to \dots$$

which means that FTP fails.
In general we have the following undecidably result.

<u>5. Theorem:</u>

It is undecidable whether a system $\mathbf{R}$ has infinite repetition free derivations.

Proof: This follows from the proof of Theorem 12 of chapter II.1.. The existence of an infinite repetition free derivation in **R** amounts to an infinite repetition free derivation in **R**(G) which is an undecidable property.

[]

As a consequence we get the undecidability of weak FTP for (f+b)-systems. Therefore a positive criterion which ensures FTP is useful.

6. Proposition:

(R_f+R_b) has weak FTP if there is some KB-ordering with a weight function ω such that $\omega(s)\geq\omega(t)$ for each $s\to t\in R_f\cup R_b$.

Proof: In a derivation the weight never increases and there are only finitely many words for each weight.

[]

The criterion in the last proposition is in particular satisfied in the examples from above where certain reductions with non-descreasing weights have been reversed. It may however for a given KB-ordering be possible to reverse also strictly weight descreasing reductions.
This is possible because some other KB-ordering may be used. The existence of such an ordering is decidable.

7. Proposition:

Suppose **R** is a finite system. Then the existence of a weight function ω for which $\omega(s) \geq \omega(t)$ holds for each $s \to t \in \mathbf{R}$ is decidable; in the positive case such ω can be effectively constructed.

<u>Proof:</u> Suppose the generators are $a_1,\dots,a_n$. Then each reduction $s \to t$ induces an inequality of the form

$$\sum_{k=1}^{n} k(s)\cdot\omega(a_k) \geq \sum_{k=1}^{n} k(t)\cdot\omega(a_k)$$

where $k(u)$ is the number of occurrences of a_k in u.
If we introduce unknowns x_k we have to solve each inequality

$$\sum_{k=1}^{n} k(s)\ x_k \geq \sum_{k=1}^{n} k(t)\ x_k$$

with extra condition $x_k \geq 1$ in the domain of the integers $\mathbf{N}$. This integer linear programming problem is a formula of Presburger arithmetic which is decidable. Presburger arithmetic, however, is not very efficient; in fact it is super-exponential. A more efficient method is Bledsoe's SUP-INF method and its refinement by Shostak (cf. [Ble] and [Sho]). This method gives for the linear programming problem numbers INF_k and SUP_k such that

$$INF_k = Min(x \mid x \text{ occurs in a real solution for } x_k)$$
$$SUP_k = Max(x \mid x \text{ occurs in a real solution for } x_k)$$

Because our problem has real solutions iff it has integer solutions it is solvable iff INF ≤ SUP. In the latter case a solution is constructed by substituting SUP_1 for x and reducing the problem in this way.

[]

We note that for deciding weak FTP we need not to know the actual solution of the linear programming problem.

III.3. The Church-Rosser Property of Forward-Backward Systems

In this chapter we will show that for forward-backward-systems CR is not decidable. We will consider a special kind of forward-backward systems (see below), and we will prove that even for this class CR is undecidable.

We assume throughout this paragraph that $\mathbf{R} = \mathbf{R}_f + \mathbf{R}_b$, and that for some weight function ω we have:

$$\text{(i)} \quad s \to t \in \mathbf{R} \quad \Rightarrow \quad \omega(s) \geq \omega(t)$$

$$\text{(ii)} \quad s \to t \in \mathbf{R} \text{ and } \omega(s) = \omega(t) \quad \Rightarrow \quad t \to s \in \mathbf{R}$$

We will use the following notation:

$$u \longmapsto v \quad \text{if} \quad u \;{}_{\mathbf{R}}\!\!\to v \quad \text{and} \quad \omega(u) > \omega(v)$$

$$u \vdash\!\!\dashv v \quad \text{if} \quad u \;{}_{\mathbf{R}}\!\!\to v \quad \text{and} \quad \omega(u) = \omega(v)$$

$$u \mapsto\!\!\!| \; v \quad \text{if} \quad u \longmapsto v \quad \text{or} \quad u \vdash\!\!\dashv v.$$

From (ii) it follows, that $\vdash\!\!\dashv$ is symmetric. Let $\vdash\!\!\dashv^*$, $\longmapsto^*$ and $\mapsto\!\!\!|^*$ describe the transitive and reflexive hulls of $\vdash\!\!\dashv$, $\longmapsto$ and $\mapsto\!\!\!|$.

$$\mathbf{R}_{\mapsto} := \{s \to t \in \mathbf{R} \mid \omega(s) > \omega(t)\}$$

$$\mathbf{R}_{\vdash\!\dashv} := \{s \vdash\!\!\dashv t \mid s \to t \in \mathbf{R} \text{ and } \omega(s) = \omega(t)\}.$$

$s \vdash\!\!\dashv t$ denotes a new kind of rules, which can be applied in both directions, i.e.:
$u \vdash\!\!\dashv v$ iff v is obtained from u by applying $s \to t$ or $t \to s$.

We put $\mathbf{M} = \Sigma^*/\vdash\!\dashv^*$. The equivalence class of a word w is denoted by [w].

We write $[u] \;_{\mathbf{M}}\!\!\longrightarrow [v]$ iff $\exists u',v'$ $u \vdash\!\dashv u'$ and $v \vdash\!\dashv v'$ and $u' \vdash\!\!\rightarrow v'$

Let $_{\mathbf{M}}\!\!\rightarrow^*$ denote the transitive and reflexive hull of $_{\mathbf{M}}\!\!\rightarrow$.

The rules $[s] \;_{\mathbf{M}}\!\!\rightarrow [t]$, $s \rightarrow t \in \mathbf{R}_{\vdash\!\rightarrow}$ give a reduction system $\mathbf{R}_{\mathbf{M}}$ on the monoid **M** which only has forward rules.

1. Lemma:

The following are equivalent:

(i) **R** has CR

(ii) $\mathbf{R}_{\mathbf{M}}$ has CR

(iii) $\mathbf{R}_{\mathbf{M}}$ has WCR

Proof: The equivalence of (i) and (ii) is left as an exercise to the reader. The equivalence of (ii) and (iii) is literally the same as the one for term algebras and the free monoids.

[]

This gives a criterion to check the completeness of **R** by testing WCR for $\mathbf{R}_{\mathbf{M}}$.

Now we come to a central but somewhat complicated technical notion:

2. Definition:

For $w', w'', l \in \Sigma^*$ the narrow derivation $w' \vdash l \rightarrow w''$ is defined as:

There is some finite double sequence of words $\tilde{w}(\mu,\nu)$, $0 \le \mu \le \nu \le n$ such that $\tilde{w}(0,0)=l$, $\tilde{w}(0,n)=w'$, $\tilde{w}(n,n)=w''$ and for all $r<n$ there are $t \vdash\!\dashv s \in R_{\vdash\!\dashv}$ and $u,v \in \Sigma^*$ such that the following holds:

(1) $(\forall \mu \le r)\ \tilde{w}(\mu,r+1) = u\tilde{w}(\mu,r)v$

(2) [(2a) $[(\exists u' \neq e)(\exists v' \in \Sigma^*)$

$t = uu'$ and $(v = e$ or $v' = e)$

and $tv' = u\tilde{w}(r,r)v = \tilde{w}(r,r+1)$

and $sv' = \tilde{w}(r+1,r+1)$]

or (2b) $[(\exists v' \neq e)(\exists u' \in \Sigma^*)$

$t = v'v$ and $(u = e$ or $u' = e)$

and $u't = u\tilde{w}(r,r)v = \tilde{w}(r,r+1)$

and $u's = \tilde{w}(r+1,r+1)$]

or (2c) $[(\exists u',v' \in \Sigma^*)$

$u = v = e$

and $u'tv' = \tilde{w}(r,r) = \tilde{w}(r,r+1)$

and $u'sv' = \tilde{w}(r+1,r+1)$]].

First we want to catch the intuitive meaning of this definition. Suppose we want to apply the rule $s_o \vdash\!\dashv t_o$ to l (in the direction $t_o \rightarrow s_o$). This might be impossible, because t_o need not to be a subword of l. But it might be possible to enlarge the word l and to apply the rule.

Take for example $t_o = u_o u_o'$, $u_o' \neq e$, u_o' a subword of l.

Then we put: $\tilde{w}(0,0) := l$

$\tilde{w}(0,1) := u_o\tilde{w}(0,0) = t_o v_o'$ $(v_o := e)$

$\tilde{w}(1,1) := s_o v_o'$

Hence we have $\tilde{w}(0,1) \vdash\!\dashv \tilde{w}(1,1)$ by $s_o \vdash\!\dashv t_o$.

The following figure shows the meaning of the conditions 2a-2c:

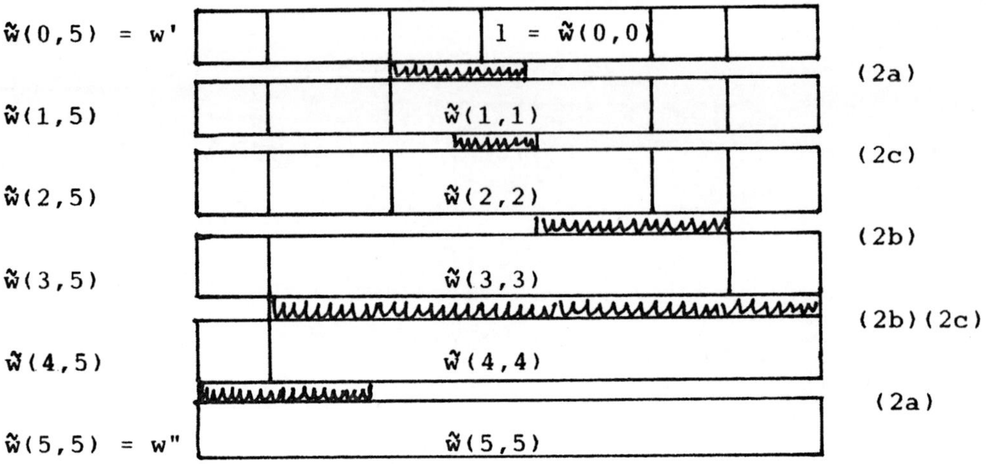

Note that in order to come from $\tilde{w}(r,r)$ to $\tilde{w}(r+1,r+1)$ by the rule $t \vdash\!\dashv s$, t must overlap with $\tilde{w}(r,r)$ but t need not to be a subword of $\tilde{w}(r,r)$.

The next definition discusses possibilities of extending the words w' and w" from above:

3. Definition:

$w_1 \vdash l_1 - l_2 \rightarrow w_2$ iff there are $u,v,w_1',w_2' \in \Sigma^*$ such that

(1) $uw_1'v = w_1$ and $uw_2'v = w_2$ and $w_1 \vdash l_1 \rightarrow w_2'$ and

(2) (2a) $(l_2 = w_2)$
or (2b) $(\exists u' \in \Sigma^*)(u = e$ and $w_2 = u'l_2)$
or (2c) $(\exists v' \in \Sigma^*)(v = e$ and $w_2 = l_2v')$
or (2d) $(\exists u',v' \in \Sigma^*)(u = v = e$ and $w_2 = u'l_2v')$

In this case we call the derivation

$w_1 = uw_1'v = u\tilde{w}(0,n)v \vdash\!\dashv u\tilde{w}(1,n)v \vdash\!\dashv \ldots \vdash\!\dashv u\tilde{w}(n,n)v = w_2$

also a <u>narrow</u> <u>derivation</u>.

The following figures show the meaning of the conditions (2a-2d):

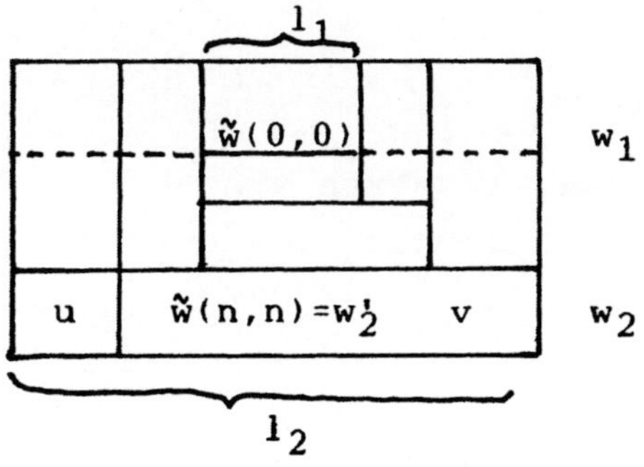

(2a)

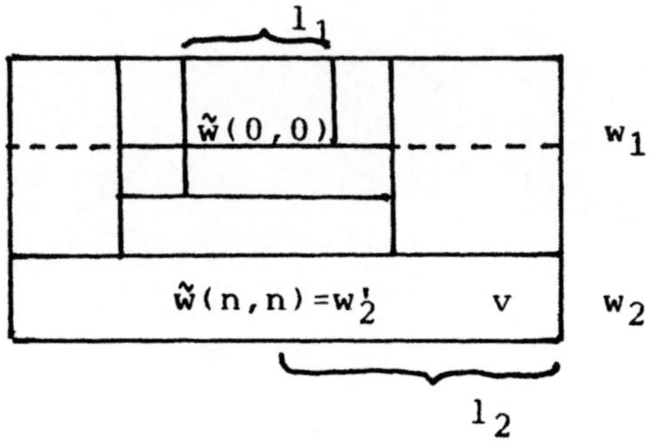

(2b)

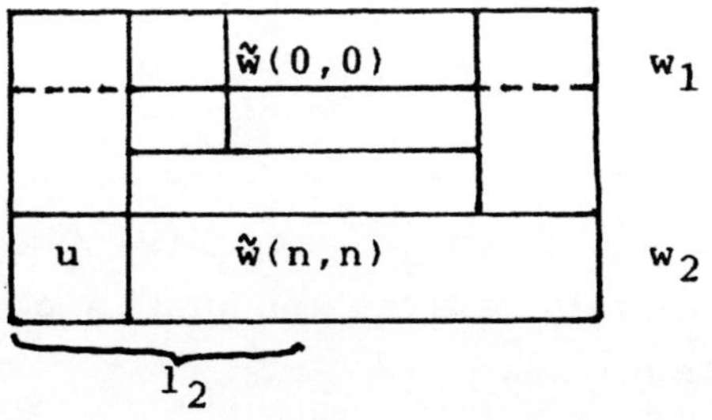

(2c)

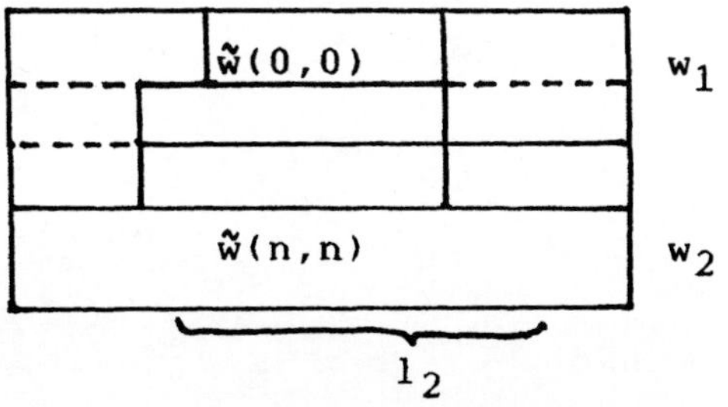

(2d)

Note, that l_2 must overlap $\tilde{w}(n,n)$, but l_2 need not be a subword of $\tilde{w}(n,n)$.

4. Lemma:

If $u_1 l_1 v_1 \vdash\!\dashv^* u_2 l_2 v_2$ then one of the following three statements holds:

(i) There are $v_2', u_2', v_1', v_1'', w \in \Sigma^*$ such that

(i1) $u_1 \vdash\!\dashv^* u_2 l_2 v_2' u_1'$ and $v_1 \vdash\!\dashv^* v_1' v_1''$

and (i2) $u_1' l_1 v_1' \vdash l_1 \rightarrow w$

and (i3) $v_2 = v_2' w v_1''$

(ii) There are $u_1'', u_1', v_1', u_2', w \in \Sigma^*$ such that

(ii1) $u_1 \vdash\!\dashv^* u_1'' u_1'$ and $v_1' u_2' l_2 v_2 \vdash\!\dashv^* v_1$

and (ii2) $u_1' l_1 v_1' \vdash l_1 \rightarrow w$

and (ii3) $u_2 = u_1'' w u_2'$

(iii) There are $u_1'', u_1', v_1', v_1'', u_2', v_2' \in \Sigma^*$ such that

(iii1) $u_1 \vdash\!\dashv^* u_1'' u_1'$ and $v_1' v_1'' \vdash\!\dashv^* v_1$

and (iii2) $u_1' l_1 v_1' \vdash l_1 - l_2 \rightarrow u_2' l_2 v_2'$

and (iii3) $u_2 = u_1'' u_2'$ and $v_2' v_1'' = v_2$

Proof: Let $u_1 l_1 v_1 = w_o \vdash\!\dashv w_1 \vdash\!\dashv \ldots \vdash\!\dashv w_{n-1} \vdash\!\dashv w_n = u_2 l_2 v_2$.

We write $u \sqsubseteq v$ if u is a subword of v.
We will try to find a double sequence $\tilde{w}$ which satisfies the conditions in Definition 2 for a narrow derivation and which satisfies $\tilde{w}(\nu,r) \sqsubseteq w_\nu$ for $\nu \leq \nu \leq n$.

We assume $\tilde{w}(\nu,r) \sqsubseteq w_\nu$ for $\nu \leq r \leq \lambda$.
Of course, there is at most one possibility to define

$$\tilde{w}(\nu,\lambda+1) \sqsubseteq w_\nu \quad \text{for } \nu \leq \lambda+1.$$

If we cannot define $\tilde{w}(\nu,\lambda+1)$ properly, then by condition 2 of Definition 2 one of the following two cases takes place:

(a) For some $\bar{w}, w', w'' \in \Sigma^*$ and some $t \mapsto s \in R_H$:
$w_\lambda = \bar{w}\tilde{w}(\lambda,\lambda)w'tw''$ and $w_{\lambda+1} = \bar{w}\tilde{w}(\lambda,\lambda)w'sw''$ holds.

(b) For some $\bar{w}, w', w'' \in \Sigma^*$ and some $t \mapsto s \in R_H$:
$w_\lambda = w'tw''\tilde{w}(\lambda,\lambda)\bar{w}$ and $w_{\lambda+1} = w'sw''\tilde{w}(\lambda,\lambda)\bar{w}$ holds.

Because both cases are similar we only consider case (a). In this case we have for $\nu \le \lambda$: $w_\nu = \bar{w}\tilde{w}(\nu,\lambda)w'tw''$.

We will make a global change and replace the sequence $w_o,\dots,w_n$ by some more appropriate sequence $\bar{w}_{-1},\bar{w}_o,\dots,\bar{w}_{n-1}$. For this purpose put

$$\bar{w}_{-1} = w_o = \bar{w}\tilde{w}(0,\lambda)w'tw''$$
$$\bar{w}_\nu := \bar{w}\tilde{w}(\nu,\lambda)w'sw'' \quad \text{for } 0\le\nu\le\lambda,$$
$$\bar{w}_\nu := w_{\nu+1} \quad \text{for } \lambda\le\nu\le n-1$$

The following figure shows the relationship between the sequences $(w_\nu)_{0\le\nu\le n}$ and $(\bar{w}_\nu)_{-1\le\nu\le n-1}$

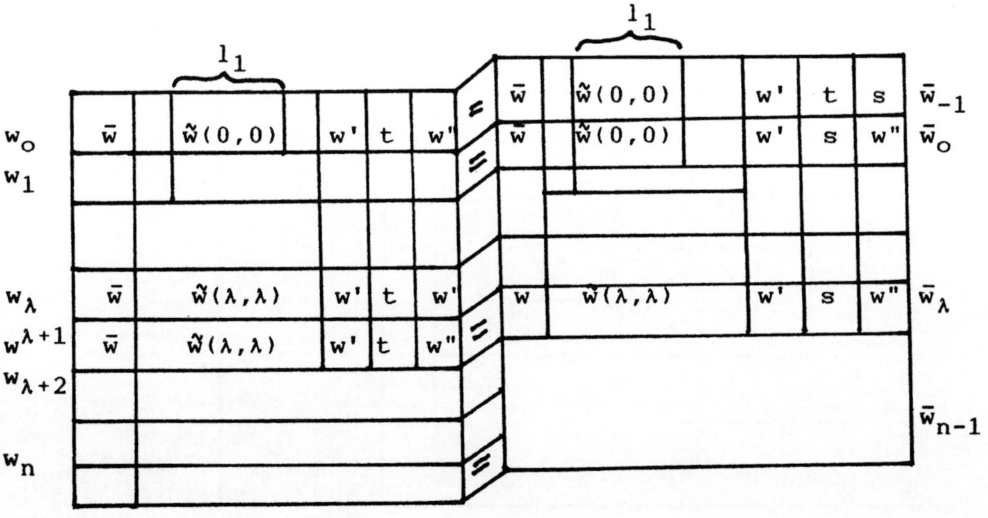

If $\tilde{w}$ is again not properly expandable (in $(\bar{w}_\nu)$), we iterate the process of changing the sequence.
Note, that before we had gone from (w_ν) to $(\bar{w}_\nu)$, we had to go through $(w_\nu)_{\lambda+1\le\nu\le n}$, but after the change we only have to

go through $(\bar{w}_\nu)_{\lambda+1\le\nu\le n-1}$. Hence the process will stop eventually after k changes of the sequence and (n-k)-times extending $\bar{w}$. Then we will have the following situation:

$$w_o = \bar{w}_{-k} \vdash\!\dashv \bar{w}_{-k+1} \vdash\!\dashv \cdots \vdash\!\dashv \bar{w}_o \vdash\!\dashv \cdots \vdash\!\dashv \bar{w}_{n-k} = w_n$$

$$w_o = \bar{u}_1''\bar{u}_1'l_1\bar{v}_1'\bar{v}_1''$$

$$\tilde{w}(0,n-k) = \bar{u}_1'l_1\bar{v}_1'$$

$$\bar{w}_\nu = \bar{u}_1''\tilde{w}(\nu,n-k)\bar{v}_1'' \quad \text{for } 0\le\nu\le n-k$$
$$\text{where } u_1 \vdash\!\dashv^* \bar{u}_1''\bar{u}_1' \text{ and } v_1 \vdash\!\dashv^* \bar{v}_1'\bar{v}_1''$$

$$u_2l_2v_2 = w_n = \bar{w}_{n-k} = \bar{u}_1''\tilde{w}(n-k,n-k)\bar{v}_1''$$

This looks as follows:

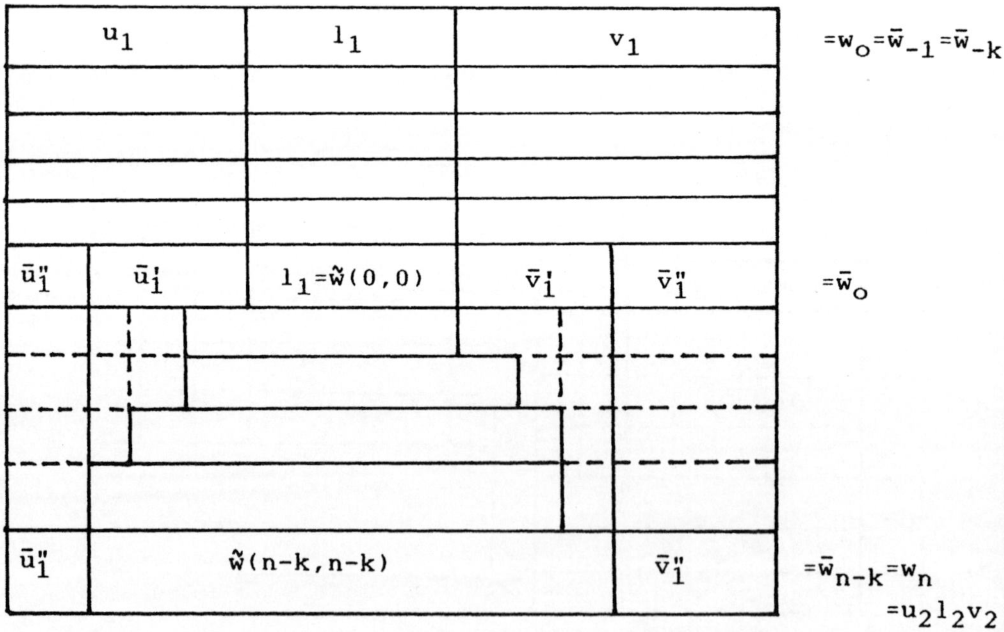

We consider the following six cases:

(I) $u_2 l_2 \subseteq \bar{u}_1''$. Then (i) holds.

(II) $l_2 v_2 \subseteq \bar{v}_1''$. Then (ii) holds.

(III) $\bar{u}_1'' \subseteq u_2$ and $\bar{v}_1'' \subseteq v_2$ and hence $l_2 \subseteq \tilde{w}(n-k,n-k)$.
Then (iii) holds.

(IV) $u_2 \subseteq \bar{u}_1'' \subsetneq u_2 l_2 \subseteq \bar{u}_1''\ \tilde{w}(n-k,n-k)$. Then $\bar{u}_1'' = u_2 u_2'$

We put $u_1'' := u_2$ $\quad$ $u_1' := u_2' \bar{u}_1'$
$w' := u_1' l_1 \bar{v}_1'$ $\quad$ $w'' := u_2' \tilde{w}(n-k,n-k)$

Then we have $w' \vdash l_1 - l_2 \rightarrow w''$ and $u_1 \longmapsto^* u_1'' u_1' = \bar{u}_1'' \bar{u}_1'$.

Hence (iii) will hold.

(V) $v_2 \subseteq \bar{v}_1'' \subsetneq l_2 v_2 \subseteq \tilde{w}(n-k,n-k)\bar{v}_1''$
This is symmetric to the above case.

(VI) $u_2 \subsetneq \bar{u}_1''$ and $v_2 \subsetneq \bar{v}_1''$ and hence $\tilde{w}(n-k,n-k) \subsetneq l_2$.

Then $\bar{u}_1'' = u_2 u_2'$ and $\bar{v}_1'' = v_2' v_2$

We put $u_1'' = u_2$, $u_1' = u_2' \bar{u}_1'$, $v_1'' = v_2$, $v_1' = v_1' v_2'$
and $w' = u_2' \tilde{w}(0,n-k) v_2' = u_1' l_1 v_1'$, $w'' = l_2$.

Then we have: $w' \vdash l_1 - l_2 \rightarrow w''$

$$u_1 \longmapsto^* u_1'' u_1' = \bar{u}_1'' \bar{u}_1'$$

$$v_1 \longmapsto^* v_1' v_1'' = \bar{v}_1' \bar{v}_1''$$

Hence (iii) will hold.

[]

The following theorem tells us the critical pairs for R_M.

5. Theorem:

R_M has WCR iff for all $l_1 \to r_1$, $l_2 \to r_2 \in R_{\mapsto}$ and for all $u_1', v_1', u_2', v_2' \in \Sigma^*$ we have:
$u_1' l_1 v_1' \vdash l_1 - l_2 \to u_2' l_2 v_2'$ implies for some $w \in \Sigma^*$
$[u_1' r_1 v_1'] \; {}_M\!\!\to^* [w]$ and $[u_2' r_2 v_2'] \; {}_M\!\!\to^* [w]$.

Proof: $\Rightarrow$: We put $a := [u_1' l_1 v_1'] = [u_2' l_2 v_2]$ and consider

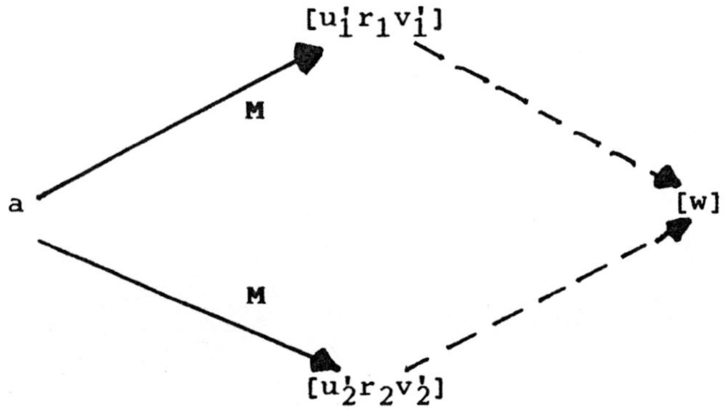

By WCR there must be some $w \in \Sigma^*$ which completes the diagram.

$\Leftarrow$: Let $a, b_1, b_2 \in M$ such that $a \; {}_M\!\!\to b_1$, $a \; {}_M\!\!\to b_2$.

Then for some $l_1 \to r_1$, $l_2 \to r_2 \in R_{\mapsto}$ and $u_1, v_1, u_2, v_2 \in \Sigma^*$ we have:

$$a = [u_1 l_1 v_1] = [u_2 l_2 v_2]$$
$$b_1 = [u_1 r_1 v_1]$$
$$b_2 = [u_2 r_2 v_2].$$

Hence, in particular $u_1 l_1 v_1 \vdash\!\dashv^* u_2 l_2 v_2$.

According to Lemma 4 we consider the following three cases:

(i) $u_1 \longmapsto^* u_2 l_2 v_2' u_1'$
$v_2 \longmapsto^* v_2' u_1' l_1 v_1' v_1'' \longmapsto^* v_2' u_1' l_1 v_1$

Hence we have $b_1 = [u_2 l_2 v_2' u_1' r_1 v_1]$ and $b_2 = [u_2 r_2 v_2' u_1' l_1 v_1]$

Hence $w := u_2 r_2 v_2' u_1' r_1 v_1$ will work.

(ii) is analogous to (i).

(iii) $u_1 \longmapsto^* u_1'' u_1'$, $v_1' v_1'' \longmapsto^* v_1$, $u_1' l_1 v_1' \vdash l_1 - l_2 \rightarrow u_2' l_2 v_2'$
with $u_2 = u_1'' u_2'$ and $v_2' v_1'' = v_2$ holds.

By our assumption there is $w' \in \Sigma^*$ such that:

$$[u_1' r_1 v_1'] \xrightarrow{}_{M}^* [w'] \quad \text{and} \quad [u_2' v_2 v_2'] \xrightarrow{}_{M}^* [w']$$

Furthermore we have

$$b_1 = [u_1 r_1 v_1] = [u_1'' u_1' r_1 v_1' v_1''] \xrightarrow{}_{M}^* [u_1'' w' v_1'']$$

and

$$b_2 = [u_2 r_2 v_2] = [u_1'' u_2' r_2 v_2' v_1''] \xrightarrow{}_{M}^* [u_1'' w v_1'']$$

Hence $w := u_1'' w' v_1''$ will work.

[]

Because there might be infinitely many critical pairs $([u_1' r_1 v_1'], [u_2' r_2 v_2'])$, the above theorem does not give us an algorithm which tests WCR. We will indeed show, that it is not decidable, whether **R** has CR. In order to do this, we will assign a reduction system to any Turing machine and its input.

6. Definition:

Let T be a Turing machine with $Q = \{q_o,\ldots,q_h\}$ the set of its states, q_o the initial state, q_h the halting state, $A = \{a_1,\ldots,a_m\}$ its alphabet $a_o = u \notin A \cup Q$.
Let $\langle,\rangle,b_1,b_2$ new symbols.
We put $\Sigma := A \cup Q \cup \{B,\langle,\rangle,b_1,b_2\}$ and denote words from Σ^* by $u,w,v,l,r,t,s,\ldots$ and words from A^+ by $\sigma,\tau\ldots$.

For $\mathbf{R} = \mathbf{R}^{(T,\tau)}$ introduce the following rules for $\tau \in A^+$:

(R1') $\langle q_o\tau\rangle \longmapsto b_1$ (R2') $\langle q_h u\rangle \longmapsto b_2$

(R1) $b_1B \vdash\!\dashv Bb_1$ (R2) $b_2B \vdash\!\dashv Bb_2$

(R3,r) $a_r\rangle B \vdash\!\dashv a_rB\rangle$ for $0\leq r\leq m$

(R4) $u\langle \vdash\!\dashv \langle u$

For every instruction $\iota = ((q_\varsigma,a_r),(q_\eta,a_\nu,\theta)) \in T$.
R will contain the following reduction:

(R5,ι,λ) $q_\varsigma a_r a_\lambda \vdash\!\dashv a_\nu q_\eta a_\lambda$ for $0\leq\lambda\leq m$, if $\theta = R$

(R6,ι,λ) $a_\lambda q_\varsigma a_r \vdash\!\dashv q_\eta a_\lambda a_\nu$ for $0\leq\lambda\leq m$, if $\theta = L$

(R7,ι) $q_\varsigma a_r \vdash\!\dashv q_\eta a_\nu$ if $\theta = N$

If the state of T is

$$\ldots\ BB\ldots Ba_{i_1}a_{i_2}\ \ldots\ \underset{\substack{\uparrow\\ q_\varsigma}}{a_{i_l}}a_{i_{l+1}}a_{i_{l+2}}\ \ldots\ a_{i_k}B\ldots\ B\ \ldots$$

then this state will be described by words of the form

$$\ldots B\langle B\ldots Ba_{i_1}a_{i_2}\ \ldots a_{i_l}q_\varsigma a_{i_{l+1}}a_{i_{l+2}}\ldots\ a_{i_k}B\ldots B\rangle B\ldots$$

7. Lemma:

If $l = \langle a_{i_1}a_{i_2}\ldots a_{i_{k'}}, q_\zeta a_{i_{k+1}}\ldots a_{i_k}\rangle$, $k>k'$ and $u \vdash l \rightarrow v$, then

$$u = B\ldots BlB\ldots B \quad \text{and}$$
$$v = B\ldots B\langle a_{j_1}\ldots a_{j_{n'}}, q_\eta a_{j_{n'+1}}\ldots a_{j_n}\rangle B\ldots B, \quad n>n'.$$

Proof: Let $\tilde{w}$ be the double sequence from Definition 2. We claim, that for $\mu \le \nu$ and for $l_\mu > l'_\mu$:

$$\tilde{w}(\mu,\nu) = B\ldots B\langle a_{i_{1,\mu}}\ldots a_{i_{l'\mu,\mu}} q_{\zeta_\mu} a_{i_{l'\mu+1,\mu}}\ldots a_{i_{l\mu,\mu}}\rangle B\ldots B,$$

This is easily shown by induction on ν; hence $u = \tilde{w}(0,n)$ and $v = \tilde{w}(n,n)$ have the desired form.

[]

8. Lemma:

Assume $l_1 = \langle a_{i_1}\ldots a_{i_{k'}}, q_\zeta\ a_{i_{k'+1}}\ldots a_{i_k}\rangle$, $k'<k$
$l_2 = \langle a_{j_1}\ldots a_{j_{n'}}, q_\eta\ a_{j_{n'+1}}\ldots a_{j_n}\rangle$, $n'<n$

Then $w_1 \vdash l_1 - l_2 \rightarrow w_2$ implies $w_1 \vdash l_1 \rightarrow w_2$.

Proof: Consider $w = uw_1'v$, $w_2 = uw_2'v$ and $w_1' \vdash l_1 \rightarrow w_2'$ By Lemma 7 we have

$w_1' = B\ldots Bl_1B\ldots B \qquad w_2' = B\ldots Bl_2'B\ldots B$
for some $l_2' = \langle a_{h_1}\ \ldots\ a_{h_{p'}}, q_\zeta\ a_{h_{p'+1}}\ \ldots a_{h_p}\rangle$, $p>p'$.

Because we cannot overlap l_2 with w_2' it follows $l_2 = l_2'$ and hence $u=v=e$, $w_1' = w_1$, $w_2' = w_2$ and finally $w_1 \vdash l_1 \rightarrow w_2$.

[]

9. Lemma:

The following are equivalent:

(a) $\exists(u_1,u_2,v_1,v_2 \in \{B\}^*)\ [u_1\langle q_o\tau\rangle v_1 \vdash\!\dashv^* u_2\langle q_h B\rangle v_2]$.
(b) The Turing machine T with input τ will stop with a blanc tape.

Proof: (b) ⇒ (a): Let p be the amount of space needed by the machine.
Putting $u_1 := B^p =: v_1$ we get

$$u_1\langle q_o\tau\rangle u_2 \vdash\!\dashv^* \langle u_1 q_o \tau v_1\rangle \qquad \text{by (R3)(R4).}$$

By (R5),(R6),(R7) and the fact that T will stop with blanc tape we see that

$$\langle u_1 q_o \tau v_1\rangle \vdash\!\dashv^* \langle u_2 q_h u v_2\rangle \quad \text{holds for some } u_2,v_2 \in \{u\}^*.$$

Again by (R3),(R4) we get

$$\langle u_2 q_h u v_2\rangle \vdash\!\dashv^* u_2\langle q_h u\rangle v_2.$$

(a) ⇒ (b): Assume $w_o = u_1\langle q_o\tau\rangle v_1$, $w_n = u_2\langle q_h u\rangle v_2$ and $w_o \vdash\!\dashv w_1 \vdash\!\dashv \dots \vdash\!\dashv w_n$.

We may assume w.l.o.g. that the length n of the derivation is minimal. Then (R4) is not used both backward and forward because in that case we would be able to find a derivation of length n-2. The same holds for (R3). Hence we may assume $w_o \vdash\!\dashv^* w_q$ and $w_q \vdash\!\dashv^* w_n$ with (R3), (R4) and $w_p \vdash\!\dashv^* w_q$ with (R5), (R6), (R7), $0 \le p \le q \le n$.
These words are of the form:

$$w_p = B\dots B\langle B\dots B q_o \tau B\dots B\rangle B\dots B$$
$$w_q = B\dots B\langle B\dots B q_h BB\dots B\rangle B\dots B\ .$$

In the last step $w_{q-1} \vdash\!\dashv w_q$ one of the rules (R5), (R6), (R7) is applied forward because T cannot enter the halting state backwards.
If one of these rules would be applied backwards then there would be a forward application followed by a backward application. But then, because T is deterministic, we would find a derivation of length n-2.
Hence in the derivation $w_p \vdash\!\dashv w_{p+1} \vdash\!\dashv \cdots \vdash\!\dashv w_q$ the rules (R5), (R6), (R7) one only applied forward. Hence this derivation describes the computation of T with input τ. We see that the machine stops with blanc tape.

[]

Now we will prove the main result of this section.

10. Theorem:

$\mathbf{R}(T,\tau)$ has CR iff the Turing machine T with input τ does not stop with blank tape.

Proof: We have to show:
$T(\tau)$ stops with blanc tape iff $\mathbf{R_M}$ has not WCR.

1) If $T(\tau)$ stops with blanc tape then for some u_1, v_1, u_2, v_2 aus $\{B\}^*$ we have by $a := [u_1 \langle q_o \tau \rangle v_1] = [u_2 \langle q_h B \rangle v_2]$ and

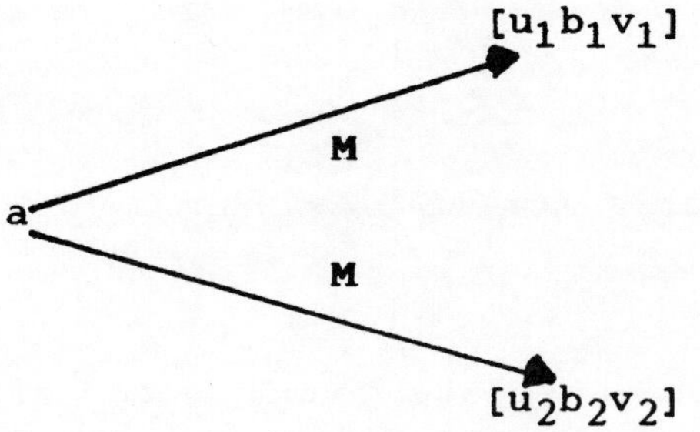

Hence R_M does not have WCR.

2) We assume that $T(\tau)$ does not stop with blanc tape. We will use Theorem 5 in order to show that R_M has WCR.

Let $l_1 \to r_1$, $l_2 \to r_2 \in R_{\mapsto}$ and $u_1 l_1 v_1 \vdash l_1 - l_2 \to u_2 l_2 v_2$.

Then $l_1, l_2 \in \{\langle q_o \tau\rangle, \langle q_h B\rangle\}$ and $u_1, u_2, v_1, v_2 \in \{B\}^*$ by Lemma 7 and Lemma 8.

We consider the following cases:

(I.) $l_1 = l_2 = \langle q_o \tau\rangle$.

Then $u_1 v_1 = u_2 v_2$, $w := b_1 u_1 v_1 = b_1 u_2 v_2$ and

$$[u_1 l_1 v_1] \underset{M}{\to} [u_1 b_1 v_1] = [b_1 u_1 v_1] = [w]$$
$$[u_2 l_2 v_2] \underset{M}{\to} [u_2 b_1 v_2] = [b_1 u_2 v_2] = [w]$$

(II.) $l_1 = l_2 = \langle q_h B\rangle$.

This can be treated like (I.).

(III.) $l_1 \neq l_2$

Without loss of generality we may assume:

$$l_1 = \langle q_o \tau\rangle \quad \text{and} \quad l_2 = \langle q_h B\rangle$$

Then we have $u_1 \langle q_o \tau\rangle v_1 \longmapsto^* u_2 \langle q_n B\rangle v_2$.

It follows from Lemma 9, that $T(\tau)$ stops with blanc tape. Hence this case is impossible.

[]

11. Theorem:

CR is undecidable for (f+b)-systems.

Proof: This is an immediate consequence of Theorem 10.

Additional remarks:

The undecidability of the CR-property for (f+b)-systems comes from the fact that finite (f+b)-systems correspond to infinite ordinary systems; in general one cannot restrict to a finite test set for WCR. As in the case of regular systems there are also here sufficient criteria which ensure that a CR-test can be carried out successfully. An example is where the narrow derivations form a regular (or even finite) set.

IV. Automata and Reductions

IV.1. General Aspects

In this chapter we consider only groups and semigroups; let Σ be the underlying alphabet.

We recall that a left regular system **R** of reductions is one where Left(**R**) is a regular set α.
In chapter III.1.2. we observed that for such **R** the set Irr(**R**) is again regular. In fact, we have Red(**R**) = $\Sigma^* \cdot \alpha \cdot \Sigma^*$ and Irr(**R**) is its complement.

Obviously Red(**R**) is closed under forming superwords and Irr(**R**) is closed under taking subwords. In order to study their corresponding automata we need some graph theoretic notions.
Suppose $A=(Q,\Sigma,\delta,q_o,F)$ is a finite deterministic automaton.

1. Definition:

(i) The word graph $\Gamma(A)$ is the following directed and labelled graph:
  (a) the vertices of $\Gamma(A)$ are the states in $Q \setminus F$;
  (b) for $p,q \in Q \setminus F$ there is an edge from p to q labelled by $a \in \Sigma$ iff $\delta(p,a) = q$.

(ii) For $\Gamma = \Gamma(A)$ the language $L(\Gamma) \in \Sigma^*$ is defined by $w \in L(\Gamma)$ iff w labels a path in $\Gamma(A)$ starting in q_o.

The closure property of Red(**R**) now implies for an automaton A which accepts Red(**R**) that Irr(**R**) = L(Γ(A)) holds. In particular, for each path in Γ(A) there is a path starting at q_o with the same labels.

<u>2. Definition:</u>

If A_o is the minimal automaton accepting Red(**R**) then Γ(**R**) = Γ(A_o) is the <u>word graph</u> for **R**.

Somewhat inconsistently we sometimes call for other A accepting Red(**R**) the graph Γ(A) also a word graph for **R**.
Each Γ(A) determines a labelled tree T(A) such that each path in Γ(A) corresponds to a path in T(A) and vice versa. T(A_o) is called the <u>word tree</u> for **R**. In T(A_o) we can associate to each node the word corresponding to a path from q_o to this node.

T(A_o) as well as Γ(A_o) is completely determined by Irr(**R**). A_o and therefore also Γ(A_o) is however not determined by the semigroup described by **R**. As we will see below there are two complete systems **R** and **R'** generated by the completion algorithm from the same group presentation using different orderings for which the minimal automata accepting Red(**R**) and Red(**R'**) are not isomorphic.

The cycle structure of the word graph Γ will play an important role for the decision problems. A simple observation is:

3. Proposition:

The following are equivalent:

(i) Irr(**R**) is finite.

(ii) The word graph has no cycles.

(iii) The word tree is finite.

The word graph and the word tree represent not only the irreducible elements but also a part of the multiplication table if **R** was complete (i.e. Irr(**R**) is in one-one correspondence to the semigroup elements). For finite Irr(**R**) the whole algebraic structure can be represented by a graph:

4. Definition:

The reduction graph G(**R**) for **R** is the labelled graph which has in addition to the nodes and edges of $T(A_o)$ the following labelled edges:

If p is an end node of $T(A_o)$ (a leave node) corresponding to the word w and a $\in \Sigma$ then there is an edge labelled with a from p to the node corresponding to Irred(wa,**R**).

Next we are interested in ways to construct these graphs directly from the reduction system which we assume is left regular.
The simplest way is to construct (from an automaton B accepting the left sides of **R**) a non-deterministic automaton A accepting Red(**R**). In case where $\mathbf{R} = \{l_i \to r_i \mid 1 \le i \le m\}$, $l_i = a_{i1} \ldots a_{iki}$, is finite this leads to the following transition graph of A:

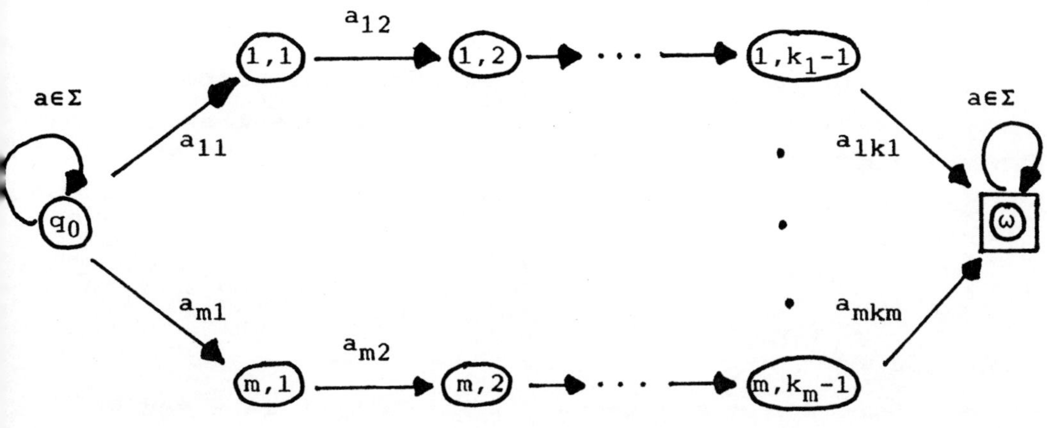

The transformation to a deterministic and minimal automaton yields the word graph.
This construction is rather inefficient, however. Therefore we will now describe a more refined method.

We recall that a word u is a proper prefix of a regular set T if for some $v \neq 1$ we have $uv \in T$; for a word u the left derivation of T was ${}_uD(T) = \{v \mid uv \in T\}$; u and w are equivalent with respect to T if ${}_uD(T) = {}_wD(T)$ and there are only finitely many non-equivalent words for each regular T.
Now suppose a left regular non-redundant set **R** of reductions is given and let L = Left(**R**) be the set of left sides of **R**. Next we choose a maximal set Q of non equivalent proper prefixes of L; hence for each proper prefix u there is some $v \in Q$ such that ${}_uD(L) = {}_vD(L)$. W.l.o.g. we can assume $1 \in Q$.

We define the following directed and labelled graph G(L):

1) Q is the set of nodes of G(L);

2) if $u \in Q$, $a \in \Sigma$, ua a proper prefix of L and ua is equivalent to $v \in Q$, then there is an edge from u to v labelled by a;

3) if $u \in Q$, $a \in \Sigma$ and $ua \in Irr(\mathbf{R})$ but ua not a proper prefix of L, then let v be the first proper prefix obtained from ua by omitting letters from left to right; v is equivalent to some $w \in Q$ and we require an edge from u to w labelled by a.

For some $w \notin Q$ we put $Q' = Q \cup \{w\}$ and add the following edges:

a) For each $u \in Q$ and $a \in \Sigma$ such that $ua \in Red(\mathbf{R})$ an edge u to w labelled by a.

b) For each $a \in \Sigma$ an edge from w to w labelled by a.

This defines an extended graph G'(L). In order to obtain the transition graph of an automaton we have to specify the initial state and the final states.
We define two automata A_1 and A_2 which have both $1 \in Q'$ as initial state by putting

(i) the set of final states of A_1 is $\{w\}$;

(ii) the set of final states of A_2 is Q.

5. Proposition:

(i) A_1 is the minimal automaton accepting $Red(\mathbf{R})$;

(ii) A_2 is the minimal automaton accepting $Irr(\mathbf{R})$;

(iii) G(L) is the word graph for **R**.

Proof: It is easy to verify the equivalence of (i), (ii) and (iii).
We will prove (i).
First we observe that ${}_vD(L) = {}_uD(L)$ implies ${}_{va}D(L) = {}_{ua}D(L)$.
Second we notice that for each $av \in Irr(\mathbf{R})$ which is not a proper prefix of L we have ${}_{av}D(L) = {}_vD(L)$.

Using induction on the length of the path this yields:
If the word v is the label of the path from 1 to $u \in Q$ then ${}_vD(L) = {}_uD(L)$.
Furthermore each $u \in Q$ is irreducible because **R** was non-redundant. For irreducible u and $a \in \Sigma$ we have $ua \in Red(\mathbf{R})$ iff ${}_{ua}D(L) = \{1\}$. This implies for a word w labelled from 1 to u:

$$wa \in Irr(\mathbf{R}) \quad \text{iff} \quad ua \in Irr(\mathbf{R}).$$

Therefore the language accepted by A_1 is Red(**R**). Because this is the automaton corresponding to the canonical right invariant equivalence relation of Red(**R**) it is the minimal one.

[]

We will illustrate this by some examples. First we consider the complete system

$$\mathbf{R} = \{B^3 \to 1,\ BA \to AB,\ A^4 \to 1\}.$$

The nondeterministic automaton accepting Red(**R**) is

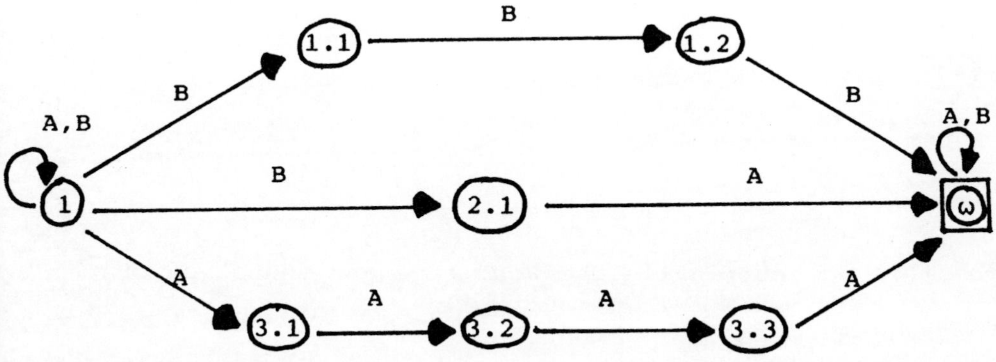

which gives as the minimal deterministic automaton for Red(**R**)

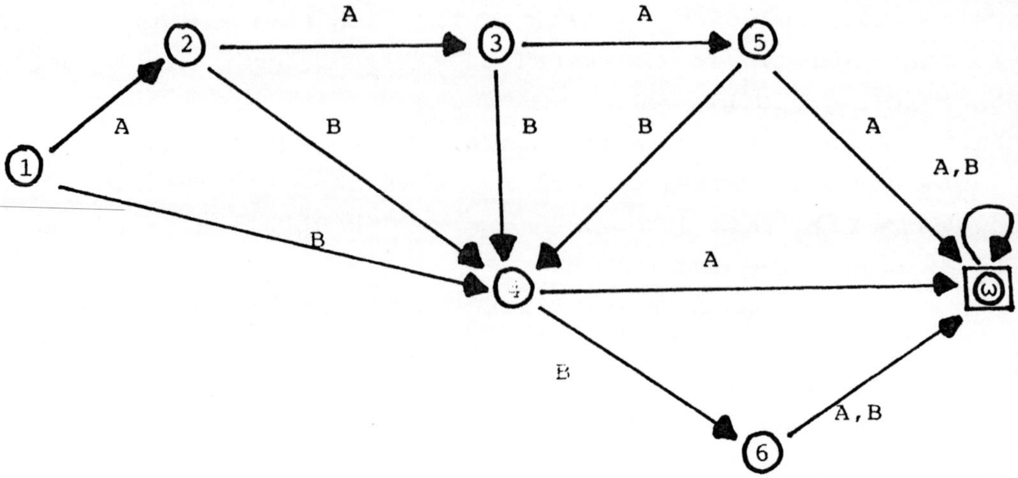

The word graph is:

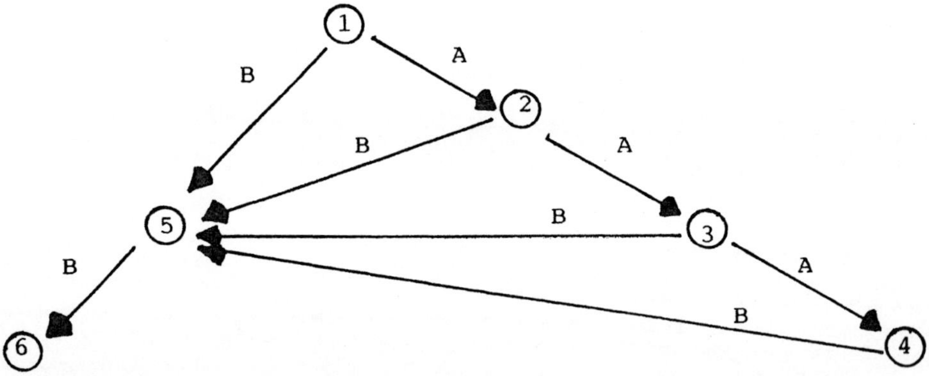

From this we get finally the word tree and reduction graph:

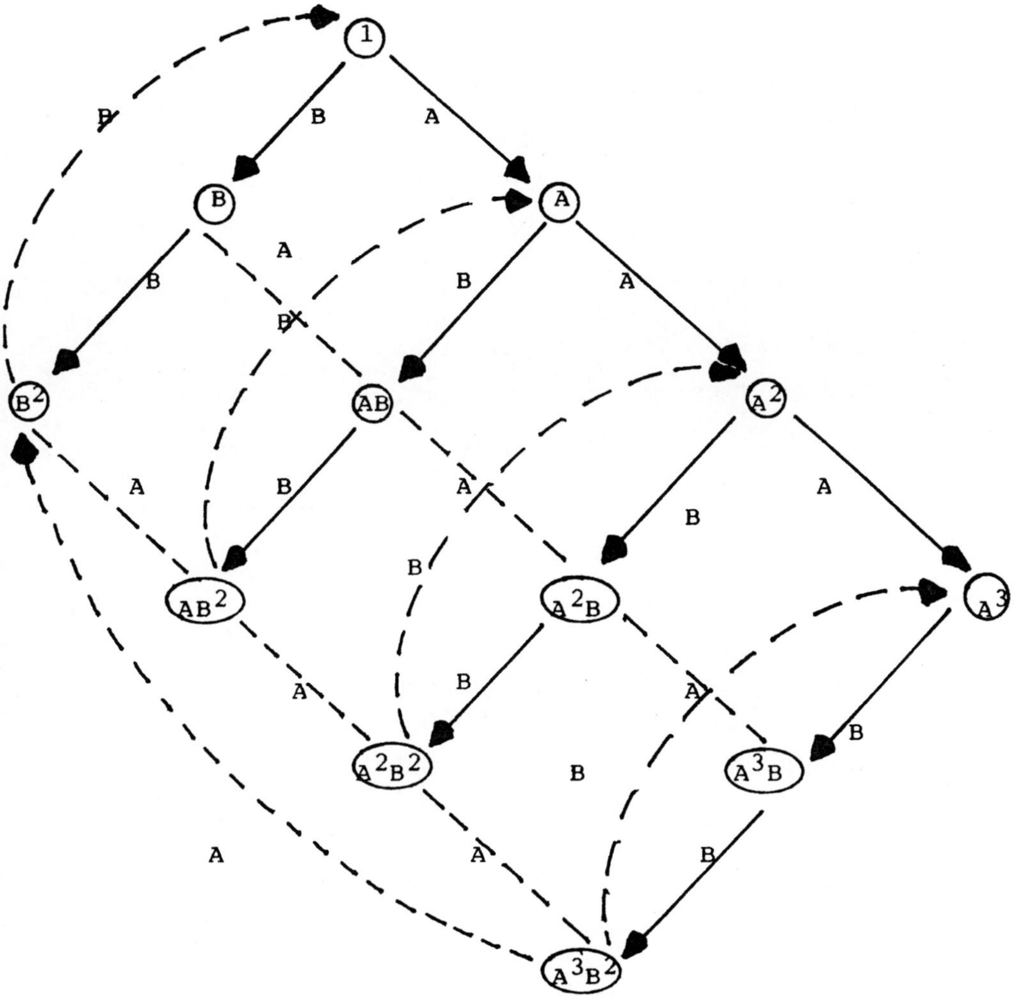

The dotted arrows are those arrows added to the word graph in order to get the reduction graph.
The proper prefixes of the left sides of **R** are

$$Q = \{1,\ B,\ B^2,\ A,\ A^2,\ A^3\}$$

which are mutually non-equivalent. This leads directly to the minimal automaton accepting Red(**R**) and to the word graph.

Since the complete system depends on the ordering used the word graph also depends on the ordering. The example of the Fibonacci group F(2,5) will show that we in fact may obtain two non-isomorphic graphs in this way.

From chapter II.4. we recall the defining equations for F(2,5)

$$ABBABA = B \quad \text{and} \quad BABABBA = A$$

The complete system for the weights $\omega(A) = 5$, $\omega(B) = 1$ leads to the word graph

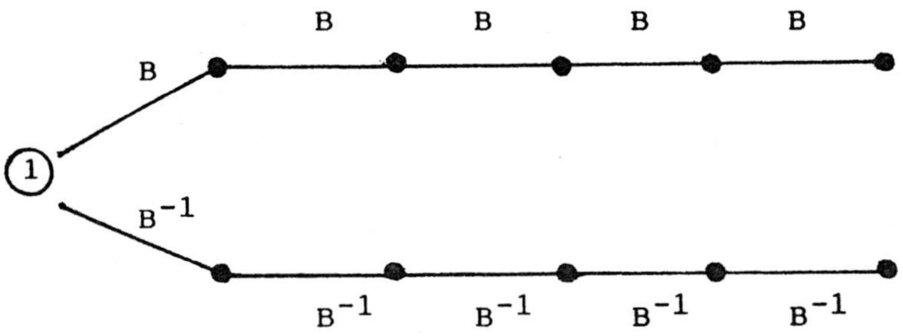

but the weights $\omega(A) = \omega(B) = 1$ induce the word graph

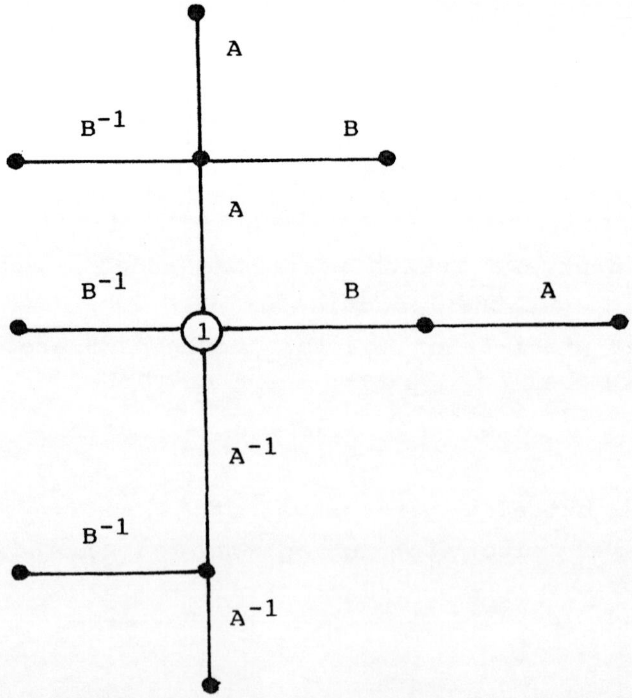

It is of interest to know how the constructions of the word graph and the word tree are related to algebraic operations on the semigroup under consideration. We consider the formation of direct products, quotients and subsemigroups.

<u>Direct products:</u> If $\mathbf{R}_i$ are complete left regular systems with word graphs $\Gamma(A_i)$ defining semigroups G_i for $1 \le i \le n$, then the word graph for the direct product $\Pi(G_i \mid 1 \le i \le n)$ is obtained by taking the disjoint union of the $\Gamma(a_i)$'s and connecting their initial nodes with some new initial node.

<u>Quotients:</u> If $\mathbf{R}$ is left regular and $u \to v$ is one new reduction, then the completion $\underline{\mathbf{R}}$ of $\mathbf{R} \cup \{u \to v\}$ need not to be left regular. This is an immediate consequence of the undecidability of the word problem for finitely presented groups and semigroups.
Because of $Irr(\underline{\mathbf{R}}) \subseteq Irr(\mathbf{R})$ the irreducible words of $\underline{\mathbf{R}}$ are represented by certain paths in the word graph $\Gamma(A)$ of $\mathbf{R}$. In case $\underline{\mathbf{R}}$ is itself left regular we observe:

a) If $Irr(\mathbf{R})$ is finite then the automaton B for $\underline{\mathbf{R}}$ is a subautomaton of the automaton A for $\mathbf{R}$ (and in this case $\underline{\mathbf{R}}$ is of course always left regular).
b) In general the automaton B for $\underline{\mathbf{R}}$ has more nodes than the one for $\mathbf{R}$ (take e.g. $\mathbf{R}$ as the trivial system describing a free monoid).
c) If there is a word u labelling a cycle in A such that $u^n \in Red(\underline{\mathbf{R}})$ for some $n>1$ then the whole cycle vanishes in B (and is replaced by a finite path, depending on n). Therefore the formation of quotients leads to a simpler cycle structure in the corresponding automata.

<u>Subalgebras:</u> If $G' \subseteq G$ is a subsemigroup of G and G is given by a left regular complete system $\mathbf{R}$ then the elements of G' are represented by certain words in $Irr(\mathbf{R})$; these words are for $G' \neq G$ not closed under the formation of sub-

words (otherwise they would contain the generators and hence all of G). If for some $u \in Irr(\mathbf{R})$ and some $n \geq 1$ the word u^n denotes a word in G' then the word graph for G' (if it exists) contains a cycle labelled by u^n.

Additional remarks:

The construction of the word graph using prefixes was first given in [Gil 79] although there was no explicit connection given to systems of reductions. A more detailed analysis can be found in [Ho 83].

IV.2. The Complexity of Reduction Algorithms

For a finite complete reduction system $\mathbf{R} = \{l_i \rightarrow r_i \mid 1 \le i \le n\}$ over a finite alphabet Σ and a word w over Σ the word in Irred(w,**R**) is unique.
In this section we deal with the complexity of computing this word.

If IRR(**R**) is finite, there is a simple linear algorithm (in the length of w). We take the word w and go letter by letter from left to right through the reduction graph, starting at the root.
The reduction graph is described in chapter IV.1.. The label of the final node is Irred(w,**R**). It is easy to see that this can be done in $O(|w|)$ steps.

This method will not work for infinite Irr(**R**) because the reduction graph is infinite, too. The rest of this section deals with this case. First of all we want to decide whether a word is irreducible or not.

1. Definition:

Let $\mathbf{R} = \{l_i \rightarrow r_i \mid 1 \le i \le n\}$ be a complete reduction system. rev(w) denotes the reversal of the word w, for example rev(abc) = cba.
The search tree of **R** consists of the paths of $\mathrm{rev}(l_i)$ with a common root and r_i as leaves.

We will illustrate this by an example and consider the complete system

$$\mathbf{R} = \{B^3 \to 1, \quad BA \to AB, \quad A^4 \to 1\}.$$

The search tree for **R** is

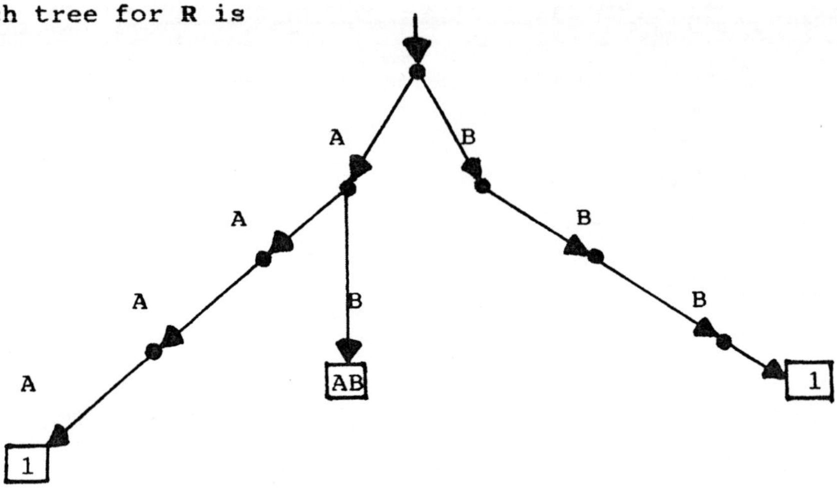

The search tree for **R** is unique because **R** is complete. So we can decide in $O(|w|)$ steps whether a given word $w \in \Sigma^*$ is irreducible; the worst-case complexity is given by $\max(|l_i|)\cdot|w|$. If w is reducible with $w' = vr_i$, then we can reduce w to $w' = vr_i$ using the search tree.

For $w \in IRR(\mathbf{R})$ and $a \in \Sigma$ we have $wa \in IRR(\mathbf{R})$ iff there is no left side $l \in \{l_i | 1 \le i \le n\}$ with $wa = xl$, x being a word over Σ.

This explains the following algorithm.

2. Reduction Algorithm:

Input of the algorithm is a finite complete reduction system $\mathbf{R} = \{l_i \rightarrow r_i \mid 1 \le i \le n\}$ and a word $w = w_1 w_2 \ldots w_{m-1} w_m$ from Σ^*.
Output of the algorithm is the unique word Irred(w,**R**).

First of all from **R** we generate the search-tree for initialization. We now describe the algorithm by simulation of two stacks in a one-dimensional array of letters.

$\$ \notin \Sigma$ is a mark to decide whether a stack is empty or not. s_1 describes the left and s_2 the right stack.
At the beginning the left stack is empty and the contents of the right stack is the word with w_1 at the top.

Let L and R be pointers for the top position, M a pointer for the "working top" (that means M is a "memory pointer" for the left stack) to remember the last position of the L pointer.

Start situation:

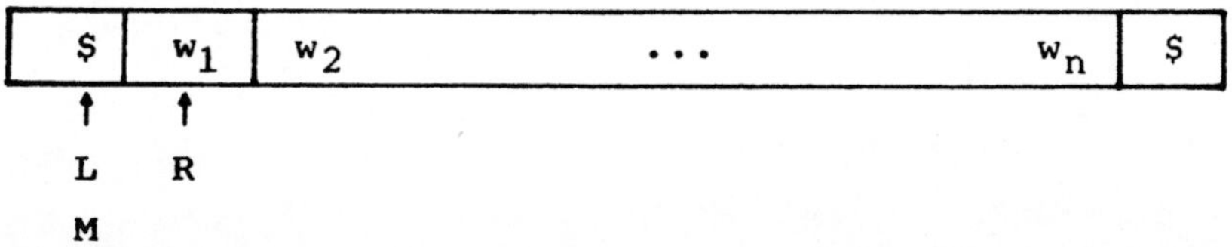

In the sequel we can describe the stacks as follows:

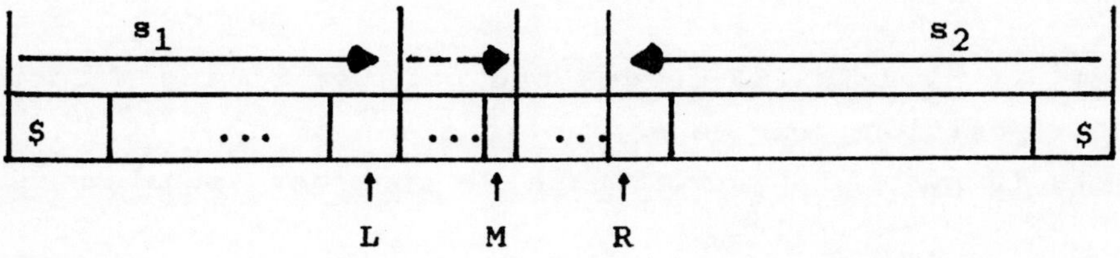

At the beginning s_1 is empty and we consider the following situation

Situation 1: (u ∈ Irr(**R**))

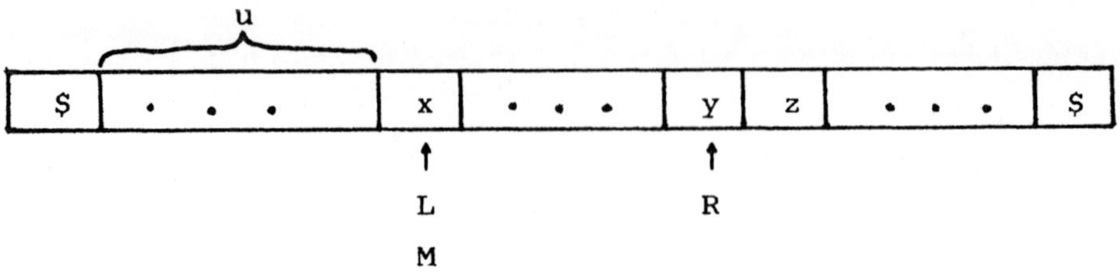

If s_2 is empty, then the algorithm terminates and {u}=Irred(w,**R**).

If s_2 is not empty then we consider situation 2, which we get directly from situation 1. We call this transformation READ, because the top letter of s_2 is transferred to the top of s_1.

Situation 2:

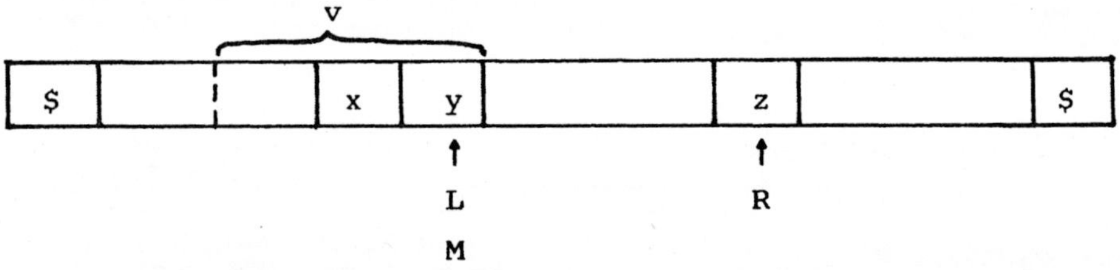

Now we test whether there is some v for which rev(v) ∈ {rev(l_i)| 1≤i≤n}. This can be done using the search tree, therefore we will call this operation SEARCH.

α) If ny ∈ IRR(**R**) we set the pointers L and M to the y position, and we get a situation of type 1.

β) If ny ∈ {l_i| 1≤i≤n} then we consider situation 3.

Situation 3:

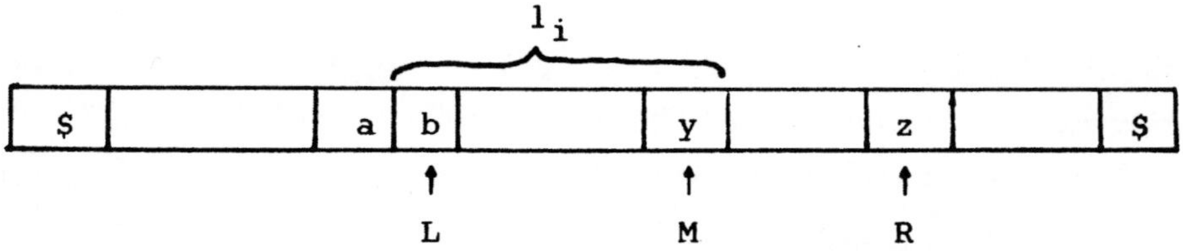

From situation 3 we get the following situation which is of type 1, we call the transformation REDUCE.

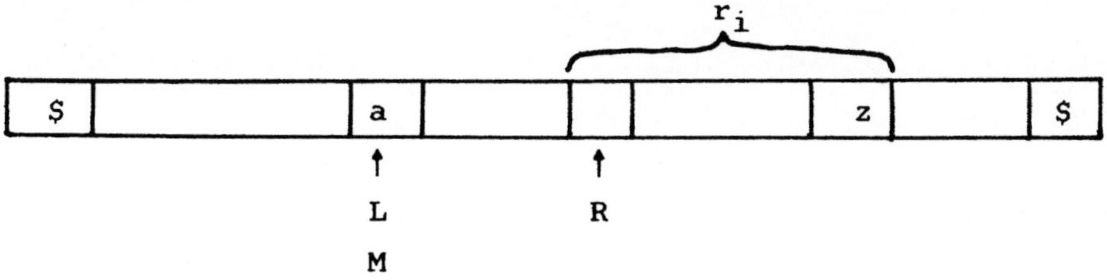

The operations are repeated until the algorithm terminates.
It is easy to see that the algorithm terminates after a finite number of steps, therefore the following proposition holds.

3. Proposition:

Let **R** be a finite complete reduction system over a finite alphabet **Σ**. The reduction algorithm computes the unique word in Irred(w,**R**) for every $w \in \Sigma^*$.

It remains to determine the complexity of the reduction algorithm. We will see that the complexity is linear in some special cases.
There also exist simple examples for which the complexity is not linear but polynomial. Unfortunately in general the complexity is exponential as we will see later.

4. Proposition:

Let $R = \{l_i \to r_i | 1 \le i \le n\}$ be a complete reduction system over the alphabet Σ for which $|l_i| > |r_i|$ for all i. The reduction algorithm computes the unique Irred(w,R) in $O(|w|)$ steps for every $w \in \Sigma^*$.

Proof: Let $l := \max\{|l_i| \,|\, 1 \le i \le n\}$ and $r := \max\{|r_i| \,|\, 1 \le i \le n\}$.

There are at most $|w|$ REDUCE-operations because for all reductions $l_i \to r_i$ we have $|l_i| > |r_i|$.

After each REDUCE step there follows a READ step and thereafter a SEARCH step. There are at most $|w| + r\cdot|w|$ READ-operations because for each application of REDUCE the stack s_2 increases by at most r letters. An upper bound for a SEARCH-operation is given if the word u is irreducible, this can be tested in at most l steps.

Hence, if T denotes the number of letter-operations we have

$$T(w) = T_{REDUCE}(w) + T_{READ}(w) + T_{SEARCH}(w)$$

$$\le r\cdot|w| + (|w|+r\cdot|w|) + l(|w|+r\cdot|w|)$$

$$= (1+2\cdot r+l+l\cdot r)|w| \in O(|w|).$$

[]

5. Definition:

A system $R = \{l_i \to r_i | 1 \le i \le n\}$ with $r_i \in \Sigma \cup \{e\}$ for all i is called monadic.

The only possible length preserving reductions for a monadic reduction system are of the form $b \to a$ with $a,b \in \Sigma$. There are at most $|\Sigma|-1$ reductions of this form because **R** is complete. From this we get directly the following corollaries.

<u>6. Corollary:</u>

Let **R** be a finite complete monadic reduction system over the alphabet Σ. The reduction algorithm computes the unique word in Irred(w,**R**) in $O(|w|)$ steps for every $w \in \Sigma^*$.

<u>7. Corollary:</u>

Let $\mathbf{R} = \{l_i \to r_i \mid 1 \le i \le n\}$ be a complete reduction system over the alphabet Σ. If for all i we have $|l_i| > |r_i|$ or $r_i \in \Sigma \cup \{e\}$ then the reduction algorithm computes the unique word in Irred(w,**R**) in $O(|w|)$ steps for every $w \in \Sigma^*$.

<u>8. Example:</u> We consider the reduction system $\mathbf{R} = \{ba \to ab\}$ over $\Sigma = \{a,b\}$ which is complete.
Hence $IRR(\mathbf{R}) = \{a^i b^j \mid i,j \in \mathbb{N}\}$.
If we reduce the word $w = b^n a^n$, $n \ge 1$, we must apply exactly $n^2 = |w|^2/_4 \in O(|w|^2)$ reductions to get the irreducible word Irred(w,**R**) $= a^n b^n$. In this case there is no restriction to the position on which we apply the reduction.

Next we consider a more complex example to show that there is not always a polynomial algorithm to reduce a word $w \in \Sigma^*$ with a finite complete reduction system.

9. Example: We consider the alphabet $\Sigma = \{*,0,1,\#,\$\}$ with order $* < 0 < 1 < \# < \$$.
It is obvious that the following set forms a complete reduction system **R**.

1:	0*	→	*1
2:	1*	→	0#
3:	#0	→	0#
4:	#1	→	1#
5:	#$	→	*$
6:	##	→	#
7:	#*	→	*

Now we reduce a word $1^n{*}\$ \in \Sigma^*$, $1^n{*}\$$ denotes a word of n 1's followed by a * and a $. f(n) denotes the number of reductions which we need to get from $1^n{*}\$$ the word $\mathrm{Irred}(1^n{*}\$,\mathbf{R}) = {*}1^n\$$.

At every reduction step exactly one of the above reductions is applicable, because the reduction process is unambiguous for words of the form $1^n{*}\$$. This is the reason why we can count the number of reductions which we need. In particular, we have

$$f(n) = 2^{n+2} - n - 4.$$

We stress the fact that the complete system in this example is also non-redundant. Exponential examples of redundant complete systems are somewhat easier to get. We also remark that in the group case no complete, non-redundant exponential system is known. We finally give two examples of further applications.

<u>10. Example:</u> Let **R** be a finite complete reduction system over the alphabet Σ.
Then there is a simple algorithm for solving the word problem.

For $w_1, w_2 \in \Sigma^*$ we first compute $u_1 = \mathrm{Irred}(w_1, \mathbf{R})$ and $u_2 = \mathrm{Irred}(w_2, \mathbf{R})$ and then we test the words u_1 and u_2 for equality. So the word problem can be solved, and the order of complexity is the same as for the reduction algorithm.

<u>11. Example:</u> Let **R** be a finite complete reduction system over the alphabet Σ.
If **R** describes a group representation and if for all $a \in \Sigma$ we can identify $a^{-1} \in \Sigma$ then we can compute for every word $w \in \Sigma^*$ the unique word $v \in \mathrm{IRR}(\mathbf{R})$ for which we have $\mathrm{Irred}(wv, \mathbf{R}) = 1$.

Let $w = w_1 w_2 \ldots w_{n-1} w_n$, $w_i \in \Sigma$, then we have $\mathrm{Irred}(ww^{-1}, \mathbf{R}) = 1$ and $w^{-1} = w_n^{-1} w_{n-1}^{-1} \ldots w_2^{-1} w_1^{-1}$. So we have $v = \mathrm{Irred}(w^{-1}, \mathbf{R})$ which we can compute by the reduction algorithm. The order of complexity is again the same.

IV.3. The Cycle Structure and the Growth Function

In chapter IV.1. we observed that for a semigroup G with a left-regular complete system finiteness is characterized by the non-existence of a cycle in the word graph Γ. In this section we will investigate the cycle structure of the word graph in more detail and see which information about G it contains. In some sense the cardinality is an asymptotic aspect of G: Infinity of G means that the formation of new words will lead to new elements in G although different words may sometimes denote the same element. We will first introduce a terminology which allows to express how many elements we get by a given number of words.

Suppose G is a semigroup generated by a finite set T. The length $l_T(g)$ of $g \in G$ with respect to T is defined as usual.

1. Definition:

The growth function γ_T: $\mathbb{N} \to \mathbb{N}$ of G with respect to T is defined by:

$$\gamma_T(n) := \mathrm{card}(\{g \in G \mid l_T(g) \leq n\})$$

An analogous definition can be given for more general algebras.

2. Examples:

a) The group of integers $(\mathbb{Z},+)$ generated by $T = \{1\}$ has the growth function $\gamma_T(n) = 2n + 1$.

b) For the free abelian groups generated by m elements we get

$$\gamma_T(n) = \sum_{p=0}^{m} 2^{l} \cdot \binom{m}{p} \cdot \binom{n}{p}.$$

c) For the free monoid over m generators we obtain

$$\gamma_T(n) \geq m^n.$$

By definition, the growth function γ depends not only on G but on the generating set T too. Certain important aspects of γ turn out to be independent of T, however.

Let $\gamma: \mathbb{N} \to \mathbb{N}$ be any function.

3. Definition:

(i) γ is <u>polynomial</u> iff $\gamma(n) \leq c \cdot n^d$ for some $c,d \in \mathbb{N}$ and all $n \in \mathbb{N}$. The least such d is called the degree of γ, denoted by $\deg(\gamma)$.

(ii) γ is <u>exponential</u> iff $\gamma(n) \geq c^n$, for some $c>1$ and all $n \in \mathbb{N}$.

Because we are interested in asymptotic aspects only we will in general require that certain inequalities hold for all $n \geq n_o$, n_o some fixed integer. This in general amounts to adding a constant summand; e.g. instead of requiring $\gamma(n) \leq c \cdot n^d$ for $n \geq n_o$ we could ask for $\gamma(n) \leq c \cdot n^d + a$ for some a and all n. In such a polynomial case c is called the head coefficient of a dominating polynomial.

We now need a few basic facts about these concepts.

4. Proposition:

Suppose two finite sets S and T generate the same semigroup G and assume $b_1 \cdot n^p \leq \gamma_T(n) \leq b_2 \cdot n^q$ for $0 \leq p \leq q$ and $b_1, b_2 > 0$.

Then there are some $c_1, c_2 > 0$ such that

$$c_1 \cdot n^p \leq \gamma_S(n) \leq c_2 \cdot n^q \qquad \text{holds.}$$

Proof: We choose k and l such that $l_T(x) \leq k$ for all $x \in S$ and $l_S(y) \leq l$ for all $y \in T$.

This implies $\gamma_S(n) \leq \gamma_T(kn) \leq b_2 \cdot k^q n^q$.

Hence the choice of $c_2 = b_2 \cdot k^q$ will work.

On the other hand we have

$$\gamma_S(ln) \geq \gamma_T(n) \geq b_1 \cdot n^p = (b_1 \cdot l^{-p}) \cdot (ln)^p.$$

Putting $d = b_1 \cdot l^{-p}$ and $n = a \cdot l + b$, $0 \leq b < l$ we get

$$\gamma_S(n) \geq \gamma_S(a \cdot l) \geq d \cdot (a \cdot l)^p = d \cdot (n-b)^p \geq c_1 \cdot$$

for suitable c_1.

[]

5. Corollary:

The properties of being polynomial resp. exponential are independent of the generating set. In the polynomial case the degree of the growth function is an invariant of the semigroup.

Hence there are three possibilities for G

a) G is polynomial
b) G is exponential
c) G is properly in between the polynomial and the exponential case.

We have seen examples for a) and b). An example for c) was reported in [vdD-Wi]; in our computational context this will turn to be impossible, however.

Of course, quotients of polynomial semigroups are again polynomial; passing to the quotient does not increase the degree.
The situation with respect to subgroups is a little more involved. We will consider the group case.

6. Proposition:

Suppose G is a polynomial group generated by S with a subgroup U. Then there are equivalent:

(i) The index [G:U] is finite
(ii) U is a finitely generated polynomial group with the same degree as G.

Proof: (i) ⇒ (ii): We start with some standard arguments. We take a finite set R of representatives of the cosets modulo U with $1 \in R$.
For each $g \in G$ we have $g = g^{*} \cdot u_g$, $g^{*} \in R$, $u_g \in U$; furthermore

(1) $u_{g_1 \ldots g_n} = u_{g_1(g_2 \ldots g_n)^{*}} \cdot \ldots \cdot u_{g_{n-1}g_n{}^{*}} \cdot u_{g_n}$ and
(2) $u_{g^{-1}r} = (u_{g(g^{-1}r)^{*}})^{-1}$ holds.

From this we see that $u \in U$ is a product of elements of the form u_{sr}, $s \in S \cup S^{-1}$, $r \in R$.
Hence u is finitely generated by $T = \{u_{sr} | s \in S, r \in R\}$ and

$$\gamma_S(n) \leq [G:U] \cdot \gamma_T(n) \qquad \text{holds.}$$

The assertion follows from the fact that $\deg(\gamma_T) \leq \deg(\gamma_S)$ is always true.

(ii) $\Rightarrow$ (i): We take finite generating sets $X \subseteq Y$ of U resp. G. Because of $\gamma_X(n) \leq \gamma_Y(n)$ and the assumption there are some $c>0$, $n_o \in \mathbb{N}$ such that

$$\gamma_Y(n) / \gamma_X(n) \leq c \qquad \text{for all } n \geq n_o.$$

Furthermore for all $\varepsilon>0$, $k \in \mathbb{N}$ there are infinitely many m such that

$$(*) \qquad \gamma_X(k+m) / \gamma_X(m) \leq 1+\varepsilon \qquad \text{holds.}$$

Otherwise $\gamma_X(k+m) / \gamma_X(m) > 1+\varepsilon$ for all sufficiently large n.
Hence $\gamma_X(r) / \gamma_X(N) \geq (1+\varepsilon)^l$ for fixed N and $l = [r-N/k]$, which gives an exponential growth for U and hence for G, too.

If there are $g_1,\dots,g_t \in G$, $g_i \neq g_j \bmod U$ for $i \neq j$ such that each $l_Y(g_i) \leq k$ for each i, then for $m \geq n_o$:

$$t \leq \gamma_Y(k+m) / \gamma_X(m) \leq c\cdot(\gamma_X(k+m) / \gamma_X(m))$$

Now (*) implies $t \leq (1+\varepsilon)\cdot c$ for each $\varepsilon>0$, therefore $t \leq c$ and finally $[G:U] \leq c$.

[]

If U has an infinite index in G then the next proposition shows that the degree strictly decreases.

7. Proposition:

Suppose G is polynomial with degree d+1 and [G:U] is infinite. Then U has degree $\leq$ d.

Proof: We take finite generating sets $x \subseteq y$ for U resp. G and representatives $r \in R$ from the left cosets modulo U such that $l_y(r)$ is minimal in rU. Furthermore we put for $n \in \mathbb{N}$:

$$G_n := \{g \in G \mid l_y(g) \leq n\}$$
$$U_n := \{g \in U \mid l_x(u) \leq n\}$$
$$R_n := G_n \cap R.$$

Then we have $G_n \supseteq \bigcup ((R_k \setminus R_{k-1}) \cdot U_{n-k} \mid 1 \leq k \leq n) \cup R_0 \cdot U_n$

Because [G:U] is infinite we get $\mathrm{card}(R_k \setminus R_{k-1}) \geq 1$; from this we obtain

$$\gamma_x(n) \geq \sum_{k=0}^{n} \gamma_x(k).$$

If $\deg(\gamma_x)=m$, then $\gamma_x(n) \geq c \cdot n^m$ for some $c>0$, therefore $\gamma_x(n) \geq c \cdot \binom{n}{m}$ holds. This gives

$$\gamma_y(n) \geq c \cdot \sum_{i=0}^{n} \binom{i}{d};$$

and the right hand term is essentially a polynomial of degree m+1 which proves the assertion.

[]

The definition of the growth function referred to the notion of length. Alternatively one can refer to some other KB-ordering $\prec$ and define in an analogue way the $\prec$-growth function replacing length by weight and all the remaining concepts introduced above.

The fact that all KB-orderings differ asymptotically very little is expressed in the next proposition. Σ denotes the underlying alphabet.

8. Proposition:

Suppose γ_1 and γ_2 are the growth functions belonging to the same semigroup presentation and two KB-orderings $\prec_1$ and $\prec_2$. Then

(i) γ_1 is polynomial iff γ_2 is polynomial and in this case $\deg(\gamma_1) = \deg(\gamma_2)$ hold s.

(ii) γ_1 is exponential iff γ_2 is exponential.

Proof: Suppose the weight functions belonging to $\prec_1$ and $\prec_2$ are ω_1 and ω_2, resp..
We put $k := \max(\omega_1(a) \mid a \in \Sigma)$.
Then for all words u we have $\omega_1(u) = k \cdot \omega_2(u)$
and therefore $\gamma_2(n) \leq \gamma_1(k \cdot n)$.

Now if γ_1 is bounded by a polynomial p then

$$\gamma_2(n) \leq p(k \cdot n) \leq p(k) \cdot p(n),$$

hence γ_2 is bounded by a polynomial of the same degree.

If $\gamma_2(n) \geq e(n)$ for some exponential function e then

$$\gamma_1(n) \geq e(n/k)$$

and therefore γ_1 is exponential. The rest follows by the symmetric argument.

[]

Next we want to generalize the notion of the growth function to arbitrary partial orderings $\prec$. This can be done if a complete (not necessarily a left regular) system exists which is compatible with $\prec$. In particular, this is the case if for a given presentation the completion algorithm continues successfully.

9. Definition:

(i) Suppose **R** is a system of reductions.
The <u>R-growth</u> <u>function</u> $\gamma = \gamma_{\mathbf{R}}: \mathbb{N} \rightarrow \mathbb{N}$ is defined by
$\gamma_{\mathbf{R}}(n) := \operatorname{card}(\{u \in \operatorname{Irr}(\mathbf{R}) \mid l(u) \leq n\})$.

(ii) If $\prec$ is a partial ordering and **E** is a set of equations defining a semigroup G such that each class of words equivalent modulo E contains a $\prec$-minimal word then the <u>$\prec$-growth</u> <u>function</u> $\gamma = \gamma_{\prec}$ is defined by
$\gamma_{\prec}(n) := \operatorname{card}(\{u \mid u \text{ minimal in } [u]_E, \; l(u) \leq n\})$.

If **R** is a complete system compatible with $\prec$ then $\gamma_{\mathbf{R}} = \gamma_{\prec}$. If $\prec$ is a KB-ordering then our notation is compatible with the one from above. Compared with the original growth function γ it follows from the definition that $\gamma_{\prec}(n) \leq \gamma(n)$ is true for all n.
Now we assume that **R** is left regular and hence possesses a word graph Γ. We will use Γ in order to compute $\gamma_{\mathbf{R}}$. Suppose Γ is a directed graph with a starting node.

10. Definition:

(i) Γ is <u>polynomial</u> iff no two cycles of Γ have a node in common.

(ii) Γ is <u>exponential</u> iff Γ is not polynomial.

Exponential graphs are those where at least two cycles intersect in some node.

11. Definition:

(i) A minimal path in Γ starts from the starting node and visits each node at most once.

(ii) For a polynomial graph the cycle number $c(\Gamma)$ is the maximal number of cycles on a minimal path.

(iii) For a polynomial graph the path number $p(\Gamma)$ is the number of minimal paths which intersect t cycles, $t=c(\Gamma)$.

Note that each path determines some minimal path in an obvious manner.

12. Definition:

If Γ is labelled then a cycle word is a word which is the label of some cycle of Γ.

13. Proposition:

Suppose **R** is a left regular (not necessarily complete) system with word graph Γ.
Then:

(i) $\gamma(\mathbf{R})$ is polynomial iff Γ is polynomial. In this case $\deg(\gamma(\mathbf{R})) = c(\Gamma)$ and for some dominating polynomial the head term is less or equal to $p(\Gamma)$.

(ii) $\gamma(\mathbf{R})$ is exponential iff Γ is exponential.

<u>Proof:</u> Assume first that Γ is polynomial. Then each path P from the starting node in Γ is uniquely determined by

1) The minimal path M determined by P.
2) The number of runs through each cycle which intersects M.

A word w labelled on a path P can then be represented as

$$w = w_1 z_1^{j_1} w_2 z_2^{j_2} \ldots w_k z_k^{j_k} w_{k+1}$$

where $w_1 \ldots w_{k+1}$ is the word on M and z_i are cycle words with $l(z_i) = n_i$. Note that $w_2 \ldots w_k$ are non-empty because Γ is polynomial.

With

$$r = \sum_{i=1}^{k+1} l(w_i)$$

we get

$$\sum_{i=1}^{k} n_i j_i = l(w) - r.$$

Let $\gamma_M^*(n)$ (resp. $\gamma_M(n)$) be the number of words on some path P which has M as its minimal path such that $l(w) = n$ (resp. $l(w) \le n$). Then we have $\gamma_M^*(n) \le \binom{n-r}{k}$ and for $r \neq 0$

$$\gamma_M(n) = \sum_{p=1}^{n} \gamma_M^*(p) \le \sum_{p=1}^{n} \binom{p-r}{k} \le \binom{n-r+1}{k} \le \binom{n}{k} \le n^k.$$

For $r = 0$ we have exactly one cycle and the same inequality holds.

The number c of minimal paths is finite; let $t = c(\Gamma)$ be the maximal number of cycles intersecting some minimal path and $C = P(\Gamma)$ be the path number of Γ.

Hence $\gamma(R)$ is polynomial with degree at most t.

On the other hand each solution for the j_i of

$$(*) \qquad \sum_{i=1}^{k} n_i j_i \leq s - r$$

gives rise to some word w in L(Γ) with $l(w) \leq s$. The construction of Γ from the minimal automaton accepting Irr(**R**) shows that different solutions of (*) produce different words in L(Γ).

Choosing $m = \max(n_i \mid 1 \leq i \leq k)$ and $s = [(n-r)/m]$ we get for sufficiently large n:

$$\gamma(n) \geq \gamma(sm + r) \geq \binom{s}{k}$$

$$\geq \frac{(\ (n-r)/m - 1\ )\cdot \ldots \cdot(\ (n-r)/m - k\ )}{k!}$$

$$\geq (\ (n-r)/(m-k) - 1\ )^k$$

$$\geq a\cdot(n-b)^k \qquad \text{for some } a>0,\ b>0$$

$$\geq d\cdot n^k \qquad \text{for some } d>0.$$

Therefore we get $\gamma(n) \geq D\cdot n^t$ with t as above and n sufficiently large. Hence $\deg(\gamma) = c(\Gamma)$ holds.

Now we assume that Γ is exponential. We choose a node q in Γ such that two vertices start at q which are initial segments of two cycles; let w_1 and w_2 be the corresponding cycle words.

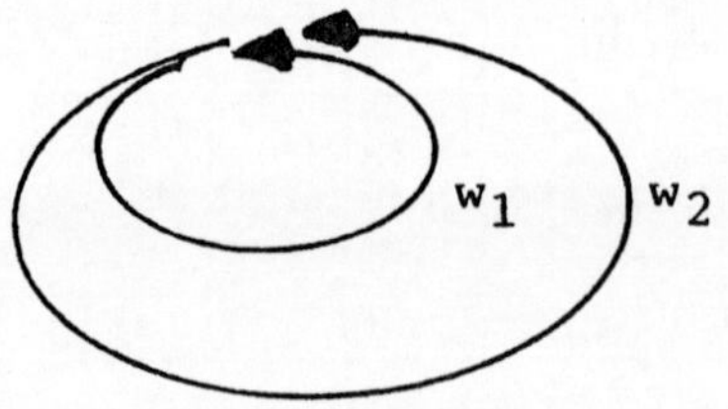

For the free semigroup SG generated by w_1 and w_2 we have $SG \subseteq L(\Gamma)$. Therefore a subset of $L(\Gamma)$ and hence $L(\Gamma)$ itself has exponential growth.
Because the cases of Γ being polynomial or exponential are exhaustive and exclusive the proposition is proved.

[]

Additional remarks:

The general connection between the growth functions and the cycle structure of the word graph as described in Proposition 13 was first observed in [Gil 79]; see also [Ho 82]. The generalization of the ordinary growth function to a $\gamma(\mathbf{R})$ is of interest if the completion algorithm does not terminate for KB-orderings. In chapter IV.4. we will investigate a group with a slower than ordinarily growing generalized growth function.

IV.4. Effective Aspects of Gromov's Theorem.

It is a natural question to characterize the class of groups with a polynomial growth function by algebraic methods. The first step in this direction was done by J. A. Wolf in [Wo].

1. Theorem:

A finitely generated nilpotent group has a polynomial growth function.

This implies that also finite extensions of nilpotent groups grow polynomially. J. Milnor conjectured in [Mi 68] that also the converse in true. The proof of this conjecture is the celebrated result of M. Gromov:

2. Theorem:

Let G be a finitely generated group. Then G has a polynomial growth function iff G has a nilpotent subgroup of finite index.

The proof uses deep analytic methods (e.g. Hilbert's 5th problem).
There is a strong interest in simplifications of this proof, in particular in removing its nonconstructive parts, (even under additional assumptions). Also one wants to find the

nilpotent subgroup "explicitly", to determine its index and to exhibit the significance of the numerical magnitudes of the growth function for the group. An important step was proved by H. Bass in [Ba 75]:

3. Theorem:

Suppose the nilpotent group has the lower central series

$$G = \Gamma_1(G) \supseteq \ldots \supseteq \Gamma_n(G) = \{e\}$$

and r_k is the rank of the free abelian group in $\Gamma_k(G)/\Gamma_{k+1}(G)$. Then G has a polynomial growth function of degree

$$d = \sum_{k=1}^{n-1} k \cdot r_k .$$

An elementary proof for the case of linear growth has also been given in [vdD-Wi].

4. Theorem:

If G is an infinite finitely generated group with growth function γ such that $\gamma(n) - \gamma(n-1) \leq n$ for some $n \in \mathbb{N}$, then G has some subgroup $U \cong \mathbb{Z}$ satisfying $[G:U] \leq n^4/_2$.

For the group theoretic concepts mentioned above we refer the reader e.g. to [Ka - Me]; some of the main notions will be recalled here, however.

The commutator of $a,b \in G$ is $[a,b] = a^{-1}b^{-1}ab$ and for subgroups $U,V \subseteq G$ the notation $[U,V]$ will mean the group generated by all $[a,b]$ with $a \in U$, $b \in V$. The lower central series of G is:

$$G = \Gamma_1(G) \supseteq \Gamma_2(G) \supseteq \ldots$$

with $\Gamma_{k+1}(G) = [\Gamma_k(G), G]$.

$\Gamma_k(G)/\Gamma_{k+1}(G)$ is in the center of $G/\Gamma_{k+1}(G)$ and G is nilpotent iff $\Gamma_n(G) = \{1\}$ for some $n \in \mathbb{N}$.

We will now connect aspects of Gromov's theorem with the word graph of the group.

Throughout this section G is finitely presented, "$\prec$" some partial term ordering on the words and $\mathbf{R}^\infty = \mathbf{R}^\infty(\prec)$ the limit system with respect to "$\prec$".
We consider the linear case.

5. Proposition:

If $\mathbf{R}^\infty$ is left regular and the $\mathbf{R}^\infty$-growth function is linear then

(i) the word graph Γ of $\mathbf{R}^\infty$ has at most two cycles;

(ii) the cycles represent a subgroup $U \cong \mathbf{Z}$ of index $[G:U] \leq (p(\Gamma)/_2)\cdot|u|$, where u is a cycle word and $p(\Gamma)$ the path number of Γ.

Proof: (i) Let u label a cycle in Γ; u generates a cyclic subgroup $U \subseteq G$ of finite index. Then there is some subgroup $U \subseteq G$ of finite index which is normal in G; U is again cyclic and generated by u^k for some k.

Now suppose v labels some cycle too. Because G/U is finite $v^n \in U$ for some $n \geq 1$, i.e. $v^n = u^l$ for some l. Therefore v is either in the given cycle or the cycle of u^{-1}, depending on $l>0$ or $l<0$.

By Proposition 10(i) of chapter IV.3. we have

$$\gamma(R^{\infty}) \leq p(\Gamma) \cdot n + d \qquad \text{for some } d;$$

the growth of the subgroup V in the generator u is 2n+1.

We want to apply the corollary of Proposition 5 of chapter IV.3.; this requires that u is a generator of G which needs not to be the case. Adding u to the generators gives an additional factor |u| such that we obtain for sufficiently large n

$$[G:U] \leq \frac{c(\Gamma) + d}{2n + 1} \cdot |u| \leq \frac{c(\Gamma)}{2} \cdot |u| .$$

[]

6. Example: $G = \langle a,b \mid a^4 = (ab)^2 = (ab^{-1})^2 = 1 \rangle$

In the KB-orderings with weights $\omega(a) > \omega(b) = 1$ the word graph is:

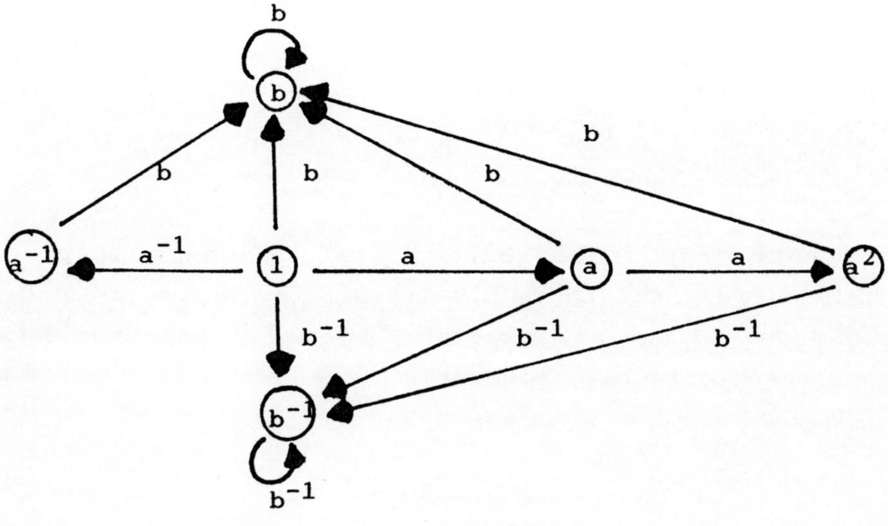

There are minimal paths intersecting a cycle and the cycle word b has length one. Therefore $[G:U] \leq {}^{8}/_{2} = 4$.

The estimate from Theorem 4 gives because of $\gamma(n)-\gamma(n-1) \leq n$ iff $n \geq 8$ the inequality $[G:U] \leq 8^4/2 = 2048$.

For the general case of an infinite group G with polynomial growth we assume Gromov's theorem; hence G has a nilpotent subgroup H of finite index. A finitely generated nilpotent group is always polycyclic (i.e. has a central series with cyclic factors) and has a torsion free subgroup $\underline{H}$ of finite index (cf. [Ka - Me]):

$$\underline{H} = H_1 \supseteq H_2 \supseteq \ldots \supseteq H_{n+1} = \{1\}, \qquad H_i/H_{i+1} \cong \mathbf{Z} \quad \text{for} \quad 1 \leq i \leq n.$$

If the limit system $\mathbf{R}^\infty$ is left regular with the word graph Γ then the elements of $\underline{H}$ are represented by a set of paths in Γ which has the same cycle structure as Γ itself because $\underline{H}$ and G have the same degree of growth.
Hence we have n cycles, labelled by certain words u_i, $1 \leq i \leq n$. Each u describing an element in $\underline{H}$ is of the form

$$u = u_1{}^{t_1(u)} \ldots u_n{}^{t_n(u)}, \quad t_i(u) \in \mathbf{Z}.$$

The elements denoted by the u_i are traditionally called a <u>Malcev-basis</u> of $\underline{H}$ and the tuples $(t_1(u),\ldots,t_n(u))$ are the <u>Malcev-coordinates</u>.

The word graph represents a part of the multiplication table, namely for those $u,v \in \mathrm{Irr}(\mathbf{R}^\infty)$ for which $uv \in \mathrm{Irr}(\mathbf{R}^\infty)$. In the above situation $\mathrm{Irred}(uv,\mathbf{R}^\infty)$ can be computed from the word graph directly, however:

7. Theorem:

For $u = u_1^{t_1(u)} \ldots u_n^{t_n(u)}$, $v = u_1^{t_1(v)} \ldots u_n^{t_n(v)}$
we have

$$\mathrm{Irred}(uv, \mathbf{R}^{\infty}) = u_1^{t_1(uv)} \ldots u_n^{t_n(uv)}$$

where

$$t_i(uv) = [\text{a polynomial over } Q \text{ in } \{t_j(u), t_j(v) \mid j<i\}] + t_i(u) + t_i(v)$$

For the proof we refer to [Ka-Me], Thm. 17, 25. The essential idea is to determine first the Malcev-coordinates of the commutators $[u_i,u_j]$ of the cycle words and to express the coordinates of more general products in these terms by using commutator identities.

In the case of quadratic growth we can make a more specific statement.

8. Proposition:

If G is nilpotent of quadratic growth then $r_1 = 2$, $r_k = 0$ for $k\geq 2$ and $C = [G,G]$ is finite.

Proof: The finiteness of the lower central series gives that for each k such that $\Gamma_k(G)$ is infinite there is some $j\geq k$ with $r(\Gamma_j(G)/\Gamma_{j+1}(G)) \geq 1$.
From the Bass formula we obtain $d=2=r_1+2r_2$ which rules out $r_1=1$.
If $r_1=2$ then $r_2=0$; in this case C has to be finite because otherwise $r_3 = r(\Gamma_2(G)/\Gamma_3(G)) \neq 0$.

Now assume $r_1=0$. Then $[G:C]$ is finite and hence C is of quadratic growth. Comparing the growth of G and C we get

$$r_2 = r(C/[C,G]) = 1, \quad r(C/[C,C]) = r_2(C) = 2.$$

On the other hand $[C,C] \subseteq [C,G]$ and both are finite; Therefore $C/[C,G]$ is the quotient of $C/[C,C]$ by a finite subgroup which is a contradiction. Hence $r_1=0$ is impossible.

[]

9. Proposition:

A group G of quadratic growth has an abelian subgroup of finite index.

Proof: We take $H \subseteq G$, $[G:H] < \infty$ and H nilpotent. Then $C = [H,H]$ is finite. On the other hand there is some $\underline{H} \subseteq H$ such that $[H:\underline{H}] < \infty$ and $\underline{H}$ is torsion free,
Therefore $C \cap \underline{H} = \{1\}$ and hence $\underline{H}$ is abelian.

[]

This result is certainly not true if the growth function has degree ≥ 3. We have, however, the following conjecture:

10. Conjecture:

If G has polynomial growth and admits a left-regular complete system for some KB-ordering, then G has an abelian subgroup of finite index.

The conjecture is supported by the results of the next section which we also think can be extended.

The V. Dyck group $D(3, 6, 2) = <a, b \mid a^3 = b^6 = (ab)^2 = 1>$ is an example of a group with quadratic growth. Choosing the weight $g(a) = g(b) = 1$ leads to the following word graph Γ:

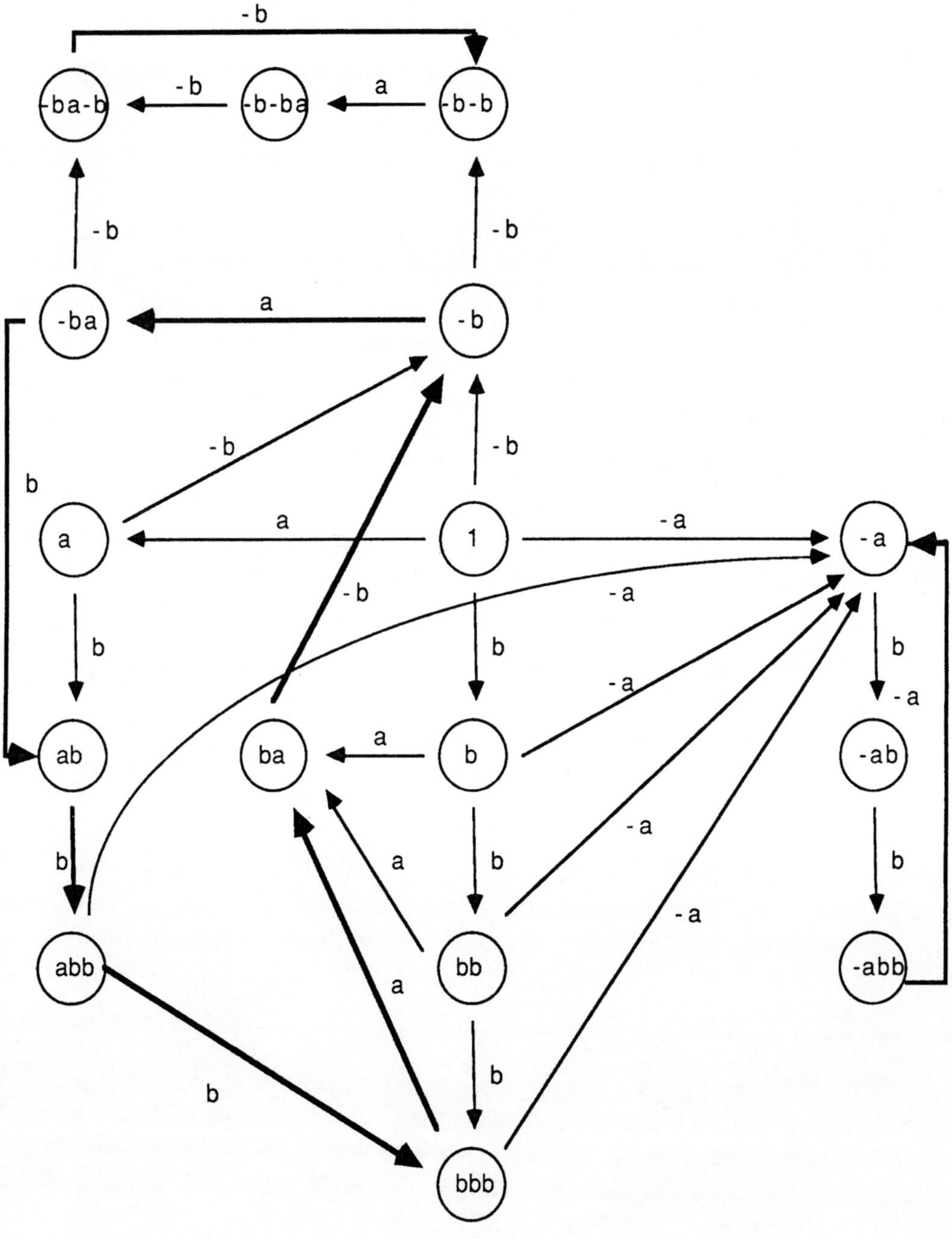

IV.5. A Relation between the Growth Function and the Completion Algorithm

If $G = \langle \Sigma, R \rangle$ is a finitely presented group and "$\prec$" some well-founded term ordering then the limit system $\mathbf{R}^\infty(\prec)$ describes some infinitary aspect of G. Another infinitary aspect is the growth function and the material in chapter IV.3. shows some relations between these aspects in case $\mathbf{R}^\infty$ is regular.
We will now use such arguments in order to prove that $\mathbf{R}^\infty$ cannot be regular in certain cases. We will restrict this discussion to the case of the free nilpotent group of class two. This group has the presentation

$$G = \langle a,b \mid [a,[b,a]] = [b,[b,a]] = 1 \rangle$$

where $[a,b] = aba^{-1}b^{-1}$ denotes as usual the commutator of a and b. It is well known that the elements of G correspond to the words $N = a^n b^m c^k$, $n,m,k \in \mathbf{Z}$, where $c = [a,b]$.

There is in fact a finite complete system $\mathbf{R}$ such that

$$N = \mathrm{Irr}(\mathbf{R}).$$

1. Definition:

$$\begin{aligned}\mathbf{R} = \{&ca \to ac,\ ca^{-1} \to a^{-1}c,\ c^{-1}a \to ac^{-1},\ c^{-1}a^{-1} \to a^{-1}c^{-1},\\ &cb \to bc,\ cb^{-1} \to b^{-1}c,\ c^{-1}b \to bc^{-1},\ c^{-1}b^{-1} \to b^{-1}c^{-1},\\ &ba \to abc^{-1},\ b^{-1}a \to ab^{-1}c,\ ba^{-1} \to a^{-1}bc,\\ &\qquad b^{-1}a^{-1} \to a^{-1}b^{-1}c^{-1}\}.\end{aligned}$$

From this definition it follows by immediate inspection that Irr(**R**) = N holds. Also, **R** has FTP and UTP; hence **R** is complete.
It is straightforward that **R** is not compatible with any KB-ordering. **R** is, however, compatible with the following partial term ordering "$\sqsubseteq$":

2. Definition:

For $u = x_1 \ldots x_n$, $x_i \in \{a, a^{-1}, b, b^{-1}, c, c^{-1}\}$, we put

$K_x(u) := \mathrm{card}(\{i \mid x_i \in \{x, x^{-1}\}\})$, $x \in \{a,b,c\}$;
$V(u) := \mathrm{card}(\{(i,j) \mid i<j,\ x_i \in \{b,b^{-1}\},\ x_j \in \{a,a^{-1}\}\})$;
$V'(u) := \mathrm{card}(\{(i,j) \mid i<j,\ x_i \in \{c,c^{-1}\},\ x_j \in \{b,b^{-1}\}\})$

and finally $u \sqsubseteq v$ iff
$(K_a(u), K_b(u), V(u), K_c(u), V'(u))$
$\leq (K_a(v), K_b(v), V(v), K_c(v), V'(v))$
where "$\leq$" denotes the lexicographic ordering on 5-tuples.

It is easy to check that "$\sqsubseteq$" is a term ordering which is compatible with **R**. In fact, taking the defining equations of G, the completion algorithm for "$\sqsubseteq$" will generate **R**.
Reducing a word this way to its normal form is a standard procedure known under the name "collecting process".
The minimal automaton A = A(**R**) accepting Irr(**R**) has a simple cycle structure, namely three cycle on a path, one for a,b and c each. Hence the cycle number is 3 and $\deg(\gamma(\mathbf{R})) = 3$, $\gamma(\mathbf{R})$ being the **R**-growth function; i.e. for the corresponding word graph $\Gamma(\mathbf{R})$ we have the cycle number $c(\Gamma(\mathbf{R})) = 3$.
In contrast to this the classical growth function of G has degree 4. An easy way to see this is to use the Bass formula (cf. chapter IV.4.):

We have $G_1 = G$, $G_2 = \langle [a,b] \rangle$ and $G_3 = \{e\}$ for the lower central series; G_1/G_2 is free abelian of rank 2 and $G_2/G_3 \cong \mathbf{N}$. This gives $d = 1 \cdot 2 + 2 \cdot 1 = 4$

Now we will investigate the behaviour of the completion algorithm for a KB-ordering. We know that this depends on the number of constants in the alphabet.
We put

$$\Sigma = \Sigma_o \cup \Sigma_1$$
$$\Sigma_o = \{a,b\}, \quad \Sigma_1 = \{c_1,...,c_n\}$$

and take as defining relations for G besides

$$[a,[b,a]] = [b,[b,a]] = 1$$

additional equations of the form

$$c_i = w_i, \quad 1 \le i \le n, \text{ for some } w_i \in (\Sigma_o \cup \Sigma_o^{-1})^*.$$

We fix an arbitrary KB-ordering "$\prec$" on Σ and denote by $\mathbf{R}^\infty = \mathbf{R}^\infty(\prec)$ the corresponding limit system.

<u>3. Proposition:</u>

Irr($\mathbf{R}^\infty$) is not regular.

Below we will give a proof for the case $\Sigma_1 = \emptyset$ and will provide some comments for the general situation. First we need some technical lemmas.

Let $\omega: \Sigma_o \to \mathbf{N}$ be the weight function defining "$\prec$" and $\alpha = \omega(a)$, $\beta = \omega(b)$.

In addition to the notions defining "$\sqsubseteq$" we put for $u=x_1...x_n$

$$K_x^+(u) := \mathrm{card}(\{i|\ x_i = x\})$$
$$K_x^-(u) := \mathrm{card}(\{i|\ x_i = x^{-1}\})$$
$$\mathbf{K}_x(u) := K_x^+(u) - K_x^-(u) \quad \text{for } x \in \Sigma_o;$$

$$V^+(u) := \mathrm{card}(\{(i,j)|\ i<j, \text{ either } x_i=b \text{ and } x_j=a \text{ or } x_i=b^{-1} \text{ and } x_j=a^{-1}\})$$
$$V^-(u) := \mathrm{card}(\{(i,j)|\ i<j, \text{ either } x_i=b \text{ and } x_j=a^{-1} \text{ or } x_i=b^{-1} \text{ and } x_j=a\})$$
$$\mathbf{V}(u) := V^+(u) - V^-(u).$$

In order to compare the two orderings we define for $r>0$

$$M(r) := \max(\mathbf{V}(u)|\ \omega(u) \le r \text{ and } \mathbf{K}_a(u) = \mathbf{K}_b(u) = 0).$$

4. Lemma:

Assume $r \ge 2(\alpha+\beta)$ and suppose u is the $\prec$-minimal word satisfying $\mathbf{V}(u) = M(r)$, $\omega(u) \le r$ and $\mathbf{K}_a(u) = \mathbf{K}_b(u) = 0$. Then u has the form $u = a^s b^{-t} a^{-s} b^t$ for some $s>0$, $t\neq 0$.

Proof: We consider $w = ab^{-1}a^{-1}b$ and get $\omega(u)\le r$, $\mathbf{V}(w)=1$. This example shows $M(r) \ge 1$ hence $\mathbf{V}(u) \ge 1$.
It follows that each of a,a^{-1},b,b^{-1} occurs in u. Therefore u is a cyclic permutation of the $M(r)$-th power of the commutator word $[a,b]$. Because $[a,b]$ is in the center of G we obtain also $u = [a,b]^{M(r)} (\mathrm{mod}\ G)$.
Next we show that n starts with some positive power of a.
Suppose this is not the case, i.e. $u = va^s z$. Because of $K_a^+(u) > 0$ we can assume $s > 0$. Then some cyclic permutation yields an $u' = u (\mathrm{mod}\ G)$ which starts with a^s; we obtain $u' \prec u$, a contradiction.

A similar argument shows that n ends with some power of b: If $u = a^s v a^k$, then $u = a^{s+k} v \pmod G$ and $a^{s+k} v \prec u$.

Therefore u now has the form

$$u = a^{s_1} b^{t_1} a^{s_2} \dots b^{t_k} \qquad \text{with} \quad \sum_{i=1}^{k} s_i = \sum_{i=1}^{k} t_i = 0.$$

We have to show k=2. Otherwise there are t_i and t_j for $i<j$ which have the same sign.
With $p = t_i$ and $q = t_j$ we get

$$u = u_1 b^p v b^q u_2 \qquad \text{with } v = av_1 = v_2 a.$$

For $m=p+q$ we consider $w_1 = u_1 b^m v u_2$ and $w_2 = u_1 v b^m u_2$.
We get $\omega(w_1) \le r$, $\omega(w_2) \le r$ and by a direct computation

$$\begin{aligned}
V(w_1) - V(u) &= V(b^m v) - V(b^p v b^q), \\
V(w_2) - V(u) &= V(v b^m) - V(b^p v b^q), \\
V(w_1) - V(u) &= m \cdot K_a(v) - p \cdot K_a(v) = q \cdot K_a(v), \\
V(w_2) - V(u) &= 0 \cdot K_a(v) - p \cdot K_a(v) = -p \cdot K_a(v).
\end{aligned}$$

There are three cases:

(i) $q \cdot K_a(v) = 0$ gives $V(u) = V(w_1) = V(w_2) = M(r)$.
Then either $b^m v \prec b^p v b^q$ and $w_1 \prec u$
or $v b^m \prec b^p v b^q$ and $w_2 \prec u$ holds
which contradicts the minimality of u.

(ii) $q \cdot K_a(v) > 0$ yields $V(w_1) > V(w)$ which is impossible.

(iii) $q \cdot K_a(v) < 0$ implies $p \cdot K_a(v) < 0$ and $V(w_2) > V(u)$; again a contradiction.

Therefore k=2 is true; the rest of the assertion follows immediately from $K_a(u) = K_b(u) = 0$.

5. Lemma:

(i) $M(4\alpha\beta n) = \alpha\beta n^2$ for all $n>0$.

(ii) If u is the $\prec$-minimal word satisfying $V(u) = M(r)$, $\omega(u) \leq 4\alpha\beta n$ and $K_a(u)=K_b(u)=0$, then u has the form

$$u = a^{-\beta n}b^{-\alpha n}a^{-\beta n}b^{-\alpha n}.$$

Proof: (i) By Lemma 4 $u = a^s b^{-t} a^{-s} b^t$ holds which gives $V(u) = s\cdot t$.
From $\omega(u) = 2\alpha s + 2\beta t \leq 4\alpha\beta$ it follows that

$$M(4\alpha\beta n) = \max\,(s\cdot t \mid s,t \in \mathbb{N},\ \alpha s + \beta t \leq 2\alpha\beta n)$$
$$\leq \max\,(x\cdot y \mid x,y \in \mathbb{R},\ x,y \geq 0,\ \alpha x + \beta y = 2\alpha\beta n) = M'.$$

If $f(x) = {}^{\alpha}/_{\beta}\cdot x\cdot(2\beta n - x)$ on the real interval $[0,2\beta n]$ then f has its maximum at $x = \beta n$ and $M' = f(\beta n) = \alpha\beta n^2$ holds. This gives the solution $x = \beta n$, $y = \alpha n \in \mathbb{N}$, therefore $M(4\alpha\beta n) = \alpha\beta n^2$ holds.

From this (ii) follows directly.

[]

Now we come back to the

Proof of Proposition 3: Suppose $Irr(R^\infty)$ is regular.
Then some finite automation A will accept $Irr(R^\infty)$, i.e. $Irr(R^\infty) = L(A)$. A pumping lemma argument applied to the class of words discussed in Lemma 5 will lead to a contradiction:
Let Q(z,u) denote the state reached from state z by input u; let z_o be the initial state of A. We choose $s_1,\dots,s_4$ sufficiently large such that the pumping lemma guarantees that for all positive integers $m_1,\dots,m_4$ the equality

$Q(z_o, a_1{}^{s_1 m_1}\ b^{-s_2 m_2}\ a^{-s_3 m_3}\ b^{s_4 m_4})$

$= \quad Q(z_o, a^{s_1}\ b^{-s_2}\ a^{-s_3}\ b^{s_4}) \quad (= z_1)$ holds.

Taking $n = s_1 \cdot s_2 \cdot s_3 \cdot s_4$ the special choice of

$$m_1 = \frac{\beta \cdot n}{s_1}, \quad m_2 = \frac{\alpha \cdot n}{s_2}, \quad m_3 = \frac{\beta \cdot n}{s_3}, \quad m_4 = \frac{\alpha \cdot n}{s_4}$$

gives a word $u = a^{\beta n}\ b^{-\alpha n}\ a^{-\beta n}\ b^{\alpha n} \in Irr(\mathbf{R}^\infty)$ because of Lemma 5; this shows that z_1 is an accepting state.

Taking t such that $x = t \cdot s_1 \cdot s_2 \ > \ y = s_2 \cdot s_4$ and choosing

$$m_1 = \frac{\beta y^2}{s_1}, \quad m_2 = \frac{\alpha y^2}{s_2}, \quad m_3 = \frac{\beta y^2}{s_3}, \quad m_4 = \frac{\alpha y^2}{s_4}$$

leads to another word $w = a^{\beta y 2}\ b^{-\alpha x 2}\ a^{-\beta y 2}\ b^{\alpha x 2}$ with $z_1 = Q(z_o, w)$.

We have $\mathbf{K}_a(w) = \mathbf{K}_b(w) = 0$ and $\mathbf{V}(w) = \alpha\beta(xy)^2$. By Lemma 5 the minimal word representing w is $a^{\beta xy} b^{-\alpha xy} a^{-\beta xy} b^{\alpha xy} \neq w$; hence $w \in Irr(\mathbf{R}^\infty)$; a contradiction.

[]

The essential part of this proof consists in an implicit discussion of the two growth functions associated with the orderings "$\sqsubseteq$" and "$\prec$". In the more general situation where finitely arbitrary many new constants can occur as abbreviations the simple use of the pumping lemma argument does not work. The only proof the authors have for this case use some sophisticated methods of nonstandard analysis which we will not discuss here.

The use of nonstandard models in the context of growth functions was first introduced in the simplified proof of Gromov's theorem given in [vdD-Wi]. In the nonstandard model the cycles in the word become essentially a real half-axis, connected by infinitesimal paths. The assumption that

$\mathbf{R}^\infty$ is regular then finally leads to a topological incompatability between $\mathbf{R}^3$ (arising from "$\sqsubseteq$") and $\mathbf{R}^4$ (arising from "$\prec$").

V. Deciding Algebraic Properties of Finitely Presented Monoids

F. Otto

Fachbereich Informatik
Universität Kaiserslautern
Postfach 3049
6750 Kaiserslautern

In this chapter we are going to investigate decision problems concerning algebraic properties of finitely presented monoids. As it will turn out, all the problems we are interested in are undecidable in general. However, we have already seen that certain problems as the word problem and the finiteness problem that are also undecidable in general, become decidable when they are restricted to monoids that are given through finite presentations involving complete reduction systems. So after establishing the undecidability results mentioned above, we will show how to solve these decision problems for finitely presented monoids which are given through certain restricted classes of complete reduction systems. In this way we will gain additional insight into the computational power of these systems.

V.1. Monoid Presentations and Tietze Transformations

Again we denote by Σ a finite alphabet, and by R a reduction system on Σ. Then in addition to the reduction relation $\xrightarrow[R]{*}$, R also induces an equivalence relation $\xleftrightarrow[R]{*}$ on Σ^*, which is defined as the smallest congruence on Σ^* containing R. As we have seen already, this congruence defines a monoid M_R, the elements of which are the congruence classes $[w]_R$ $(w \in \Sigma^*)$ and the operation of which is given by $[u]_R \circ [v]_R = [uv]_R$ $(u,v \in \Sigma^*)$, $[e]_R$ serving as the identity. Thus, M_R is the factor monoid of the free monoid Σ^* generated by Σ modulo the congruence $\xleftrightarrow[R]{*}$.

Whenever a monoid M happens to be isomorphic to M_R $(M \cong M_R)$, the ordered pair $(\Sigma;R)$ is called a <u>(monoid) presentation</u> of M with Σ being the <u>set of generators</u> and R being the <u>set of defining relations</u> of this presentation. In the following we will only be dealing with monoids that are <u>finitely presented</u>, i.e., we will only be considering monoids that are given through presentations of the form $(\Sigma;R)$, where both the sets Σ and R are finite.

Let $(\Sigma;R)$ be a finite presentation of a monoid M. Then the word problem for this presentation is the following decision problem:

INSTANCE: Two words $u,v \in \Sigma^*$.

QUESTION: Do u and v represent the same element of M, i.e., does $u \overset{*}{\underset{R}{\leftrightarrow}} v$ hold ?

So the word problem for the presentation $(\Sigma;R)$ is nothing but the word problem for the reduction system R. As is well-known there exist finite reduction systems with undecidable word problem, and hence, there exist finite monoid presentations with undecidable word problem ([Da 58]). On the other hand, different finite presentations may describe the same monoid, as the following example shows.

Example 1.1. Let $\Sigma_1 = \{a,b\}$, $R_1 = \{(a^3,e),(b^2,e),(ab,ba)\}$, $\Sigma_2 = \{b,c\}$, and $R_2 = \{(cbcbcb,e),(b^2,e),(bcb,c)\}$. Then the mapping φ defined through $\varphi(a) = cb$ and $\varphi(b) = b$ induces an isomorphism from the monoid M_{R_1} onto the monoid M_{R_2}, i.e., $(\Sigma_1;R_1)$ and $(\Sigma_2;R_2)$ are different finite presentations of the same monoid.

Obviously, this observation raises the following question: Does the decidability of the word problem depend on the actually chosen finite monoid presentation, or is it a property of the monoid presented ? In order to answer this question, but also for future reference, we introduce the notion of elementary Tietze transformation for finite monoid presentations ([Ti 08]).

Definition 1.2. Let $(\Sigma;R)$ and $(\Sigma';R')$ be two finite presentations. Then presentation $(\Sigma';R')$ is said to be obtainable from $(\Sigma;R)$ by an application of an elementary Tietze transformation of type i for some $i \in \{1,2,3,4\}$, if $(\Sigma';R')$ satisfies condition (i) given below.

(1) $\Sigma' = \Sigma$, and $R' = R \cup \{(u,v)\}$, where $u,v \in \Sigma^*$ satisfy $(u,v) \notin R$, but $u \overset{*}{\underset{R}{\leftrightarrow}} v$.

(2) $\Sigma' = \Sigma$, and $R' = R-\{(u,v)\}$, where $(u,v) \in R$ satisfies $u \overset{*}{\underset{R'}{\leftrightarrow}} v$.

(3) $\Sigma' = \Sigma \cup \{a\}$ for some letter $a \notin \Sigma$, and $R' = R \cup \{(u,a)\}$ for some word $u \in \Sigma^*$.

(4) There exist a letter $a \in \Sigma$ and a word $u \in (\Sigma-\{a\})^*$ such that $(u,a) \in R$. Let $\varphi: \Sigma^* \to (\Sigma-\{a\})^*$ denote the homomorphism induced by $\varphi(a) = u$ and $\varphi(b) = b$ for all letters $b \in \Sigma-\{a\}$. Then $\Sigma' = \Sigma-\{a\}$, and $R' = \{(\varphi(\ell),\varphi(r)) \mid (\ell,r) \in R-\{(u,a)\}\}$.

The following lemma shows that by applying elementary Tietze transformations to a finite presentation $(\Sigma;R)$ of a monoid M we only get further presentations of M. The proof of this lemma is straightforward, and therefore it is left to the reader.

Lemma 1.3. Let $(\Sigma;R)$ and $(\Sigma';R')$ be two finite presentations such that $(\Sigma';R')$ is obtainable from $(\Sigma;R)$ by an application of an elementary Tietze transformation. Then these two presentations define the same monoid, i.e., $M_R \cong M_{R'}$.

Observe that the elementary Tietze transformations of type 1 and 2 are inverses of each other, that the inverse of an elementary Tietze transformation of type 3 is one of type 4, and that the effect of an elementary Tietze transformation of type 4 on a finite monoid presentation can be reversed by a finite number of applications of elementary Tietze transformations of types 1 to 3. We will use this observation in the proof of the following theorem, which describes the basic reason for considering Tietze transformations.

Theorem 1.4. Let $(\Sigma;R)$ and $(\Sigma';R')$ be two finite presentations of the same monoid. Then there exists a finite sequence of elementary Tietze transformations that transforms $(\Sigma;R)$ into $(\Sigma';R')$.

Proof. Without loss of generality we may assume that the sets Σ and Σ' of generators are disjoint. Since $(\Sigma;R)$ and $(\Sigma';R')$ define the same monoid, we have $M_R \cong M_{R'}$. Thus, for each $a \in \Sigma$, there exists a word $u_a \in \Sigma'^*$ such that a and u_a describe the same element of this monoid. Also, for each $b \in \Sigma'$, there exists a word $v_b \in \Sigma^*$ such that b and v_b describe the same element. Using these words the presentation $(\Sigma;R)$ is transformed by a finite sequence of elementary Tietze transformations as follows:

(a) $(\Sigma;R) \rightarrow (\Sigma \cup \Sigma';R \cup \{(v_b,b) \mid b \in \Sigma'\})$ by $|\Sigma'|$ elementary Tietze transformations of type 3.

(b) Let $R_o = R \cup \{(v_b,b) \mid b \in \Sigma'\}$, and let g denote the isomorphism from $M_{R'}$ onto M_R that is induced by mapping b onto v_b for all $b \in \Sigma'$. Then for all $(\ell,r) \in R'$, $g(\ell) \overset{*}{\underset{R}{\leftrightarrow}} g(r)$ implying $\ell \overset{*}{\underset{R_o}{\leftrightarrow}} r$. Thus, $(\Sigma \cup \Sigma';R_o) \rightarrow (\Sigma \cup \Sigma';R_o \cup R')$ by $|R'|$ elementary Tietze transformations of type 1.

(c) Since for each $a \in \Sigma$, a, u_a, and $g(u_a)$ all define the same element of the monoid M_R, we have $a \overset{*}{\underset{R}{\leftrightarrow}} g(u_a) \overset{*}{\underset{R_o}{\leftrightarrow}} u_a$. Thus, we obtain $(\Sigma \cup \Sigma';R_o \cup R') \rightarrow (\Sigma \cup \Sigma';R \cup R' \cup \{(v_b,b) \mid b \in \Sigma'\} \cup \{(u_a,a) \mid a \in \Sigma\})$ by $|\Sigma|$ elementary Tietze transformations of type 1. Let $\Sigma'' = \Sigma \cup \Sigma'$ and $R'' = R \cup R' \cup \{(v_b,b) \mid b \in \Sigma'\} \cup \{(u_a,a) \mid a \in \Sigma\}$. Then $(\Sigma;R)$ has been transformed into $(\Sigma'';R'')$ by a finite sequence of elementary Tietze transformations.

(d) In an analogous manner $(\Sigma';R')$ can be transformed into $(\Sigma'';R'')$, so by the remark proceeding the theorem $(\Sigma'';R'')$ can be transformed into $(\Sigma';R')$ by a finite sequence of elementary Tietze transformations. □

It should be pointed out that the above construction does not yield a uniform process for transforming two finite presentations of the same monoid into each other, since the words u_a $(a \in \Sigma)$ and v_b $(b \in \Sigma')$ are not known in general. In particular, it does not give a solution to the isomorphism problem:

INSTANCE: Two finite presentations $(\Sigma;R)$ and $(\Sigma';R')$.

QUESTION: Do these presentations describe the same monoid, i.e., does $M_R \cong M_{R'}$ hold ?

On the other hand, it can be seen easily, that if a presentation $(\Sigma';R')$ is obtained from a finite presentation $(\Sigma;R)$ by a single application of an elementary Tietze transformation, then a solution to the word problem for $(\Sigma;R)$ also induces a solution to the word problem for $(\Sigma';R')$, and vice versa. Hence, from Theorem 1.4 we can immediately conclude the following result.

Corollary 1.5. Let $(\Sigma;R)$ and $(\Sigma';R')$ be two finite presentations of the same monoid. Then the word problem for $(\Sigma;R)$ is decidable if and only if the word problem for $(\Sigma';R')$ is decidable.

Thus, the decidability of the word problem is an invariant of finite presentations. Hence, we can speak about the decidability or undecidability of the word problem for a finitely presented monoid M. In particular, there exist finitely presented monoids with undecidable word problem.

V.2. Markov Properties of Finitely Presented Monoids

Given a monoid M through some finite presentation, one would often like to determine some of the algebraic properties of M. Here, we are specifically interested in the following decision problems:

INSTANCE: A finite presentation $(\Sigma;R)$.

1. QUESTION: Is the monoid M_R given through this presentation trivial, i.e., does $M_R \cong \{e\}$ hold ?

2. QUESTION: Is the monoid M_R finite ?
3. QUESTION: Is the monoid M_R commutative ?
4. QUESTION: Is the monoid M_R cancellative ?
5. QUESTION: Is M_R a free monoid ?
6. QUESTION: Is the monoid M_R a group ?
7. QUESTION: Does the monoid M_R contain any non-trivial idempotents?
8. QUESTION: Does the monoid M_R contain any non-trivial elements of finite order ?
9. QUESTION: Does the monoid M_R contain an element of infinite order ?
10. QUESTION: Is the monoid M_R a free group ?

In this section we will learn about a fundamental result of Markov ([Ma 51], c.f., e.g., [Mo 52]), which can be used to show that all the decision problems stated above are undecidable in general. This will then leave us the task of solving these problems for monoids that are presented by finite complete reduction systems or certain specializations thereof.

Definition 2.1. (a) A property P of monoids is called invariant if every monoid that is isomorphic to a monoid possessing property P itself possesses this property.
(b) A property P of finitely presented monoids is a Markov property, if it satisfies the following three conditions:
(0) P is invariant.
(1) There exists a finitely presented monoid M_1 which does not have property P, and which is not isomorphic to a submonoid of any finitely presented monoid having property P.
(2) There exists a finitely presented monoid M_2 having property P.
(c) Finally, a property P of finitely presented monoids is called hereditary if whenever a finitely presented monoid M has P, all finitely presented submonoids of M also have P.

Whenever P is an hereditary property of finitely presented monoids, then for checking of whether or not P is a Markov property, condition (1) given above can be relaxed to the following condition:

(1') There exists a finitely presented monoid M_1 not having property P.

Before stating and proving Markov's result we want to give some examples of Markov properties.

<u>Definition 2.2</u>. In what follows let M be a finitely presented monoid. Then the properties $P_1, P_2, \ldots, P_{10}$ are defined as follows:

(a) $P_1(M) \Longleftrightarrow$ M is <u>trivial</u>, i.e., $M \cong \{e\}$;

(b) $P_2(M) \Longleftrightarrow$ M is finite;

(c) $P_3(M) \Longleftrightarrow$ M is <u>commutative</u>, i.e., for all $m_1, m_2 \in M$, we have $m_1 \circ m_2 = m_2 \circ m_1$, where $\circ$ denotes the operation of M;

(d) $P_4(M) \Longleftrightarrow$ M is <u>cancellative</u>, i.e., for all $m_1, m_2, m_3 \in M$, $m_1 \circ m_2 = m_1 \circ m_3$ implies $m_2 = m_3$, and $m_1 \circ m_3 = m_2 \circ m_3$ implies $m_1 = m_2$;

(e) $P_5(M) \Longleftrightarrow$ M is a free monoid, i.e., $M \cong \Sigma^*$ for some finite alphabet Σ;

(f) $P_6(M) \Longleftrightarrow$ M is a group, i.e., for all $m \in M$, there is an element $m' \in M$ such that $m \circ m' = e_M$, where e_M denotes the identity of M;

(g) $P_7(M) \Longleftrightarrow$ M does not contain a <u>non-trivial idempotent</u>, i.e., there is no $m \in M$ satisfying $m \neq e_M$ and $m \circ m = m$;

(h) $P_8(M) \Longleftrightarrow$ M does not contain a <u>non-trivial element of finite order</u>, i.e., there is no $m \in M-\{e_M\}$ such that there exist $k > 1$ and $n > 0$ satisfying $m^{k+n} = m^n$, where m^i stands for $m \circ m \circ \ldots \circ m$ (i-times);

(i) $P_9(M) \Longleftrightarrow$ M does not contain an <u>element of infinite order</u>, i.e., there is no $m \in M$ satisfying $m^i \neq m^j$ for all $i,j \in \mathbb{N}$ with $i \neq j$;

(j) $P_{10}(M) \Longleftrightarrow$ M is a free group.

Obviously, all the properties $P_1, P_2, \ldots, P_{10}$ are invariant. As can be seen easily, properties P_1 to P_4 and P_7 to P_9 are hereditary, and they satisfy conditions (1') and (2), i.e., they are Markov properties. However, the remaining properties P_5, P_6, and P_{10} are not hereditary as shown by the following examples.

<u>Examples 2.3</u>. (a) Let $\Sigma = \{a,b\}$, and let $A = \{ab, aba, bab\}$. Then the monoid $M = \Sigma^*$ is a free monoid, while its submonoid $\langle A \rangle$ generated by A is not a free monoid. This can be seen from the fact that each set of words that generates $\langle A \rangle$ must contain A.
(b) Let $\Sigma = \{a, \bar{a}\}$, and let $R = \{(a\bar{a}, e), (\bar{a}a, e)\}$. Then the monoid M_R presented by $(\Sigma; R)$ is the free group F_1 of rank 1. Now consider the submonoid $\langle a \rangle$ of M_R generated by a. This submonoid is isomorphic to the free monoid $\{a\}^*$, and hence, it is not a group, let alone a free group. □

So in order to prove that properties P_5, P_6, and P_{10} are in fact Markov properties, we must check conditions (1) and (2) of Definition 2.1. Obviously, they all satisfy condition (2), which leaves us with condition (1).

Lemma 2.4. There exists a finitely presented monoid M_1 that is not isomorphic to a submonoid of any finitely generated free monoid.

Proof. Each finitely generated submonoid of a free monoid has a decidable word problem. Now let M_1 be a finitely presented monoid with an undecidable word problem. Since the undecidability of the word problem is an invariant property, we conclude that M_1 is not isomorphic to a submonoid of any finitely generated free monoid. □

So property P_5 is also a Markov property.

Lemma 2.5. There exists a finitely presented monoid M_1 that is not isomorphic to a submonoid of any group.

Proof. Let M_1 be given through the presentation $(\{a,b,c\};\{(ab,ac)\})$. Then M_1 is not cancellative, and hence, it is not isomorphic to a submonoid of any group. □

This implies that also properties P_6 and P_{10} are Markov properties. We can summarize the results obtained so far as follows.

Theorem 2.6. Properties P_1 to P_{10} are Markov properties of finitely presented monoids.

Now the main result of Markov [Ma 51] states the following undecidability result.

Theorem 2.7 ([Ma 51]). Let P be a Markov property. Then the following problem is undecidable:

INSTANCE: A finite presentation $(\Sigma;R)$.

QUESTION: Does the monoid M_R given through this presentation have property P ?

Because of Theorem 2.6 this immediately gives the following.

Corollary 2.8. The decision problems 1 to 10 are all undecidable in general.

Since a monoid M is trivial if and only if it can be presented by $(\{a\};\{(a,e)\})$, the undecidability of problem 1 induces the undecidability of the isomorphism problem. Hence, there is no hope for extending Theorem 1.4 to a solution for this problem.

It remains to prove Theorem 2.7. For doing so we need the following lemma the proof of which is essentially taken from [Bo 84].

Lemma 2.9. Let Σ be a finite alphabet, let $c,d \notin \Sigma$ be two additional letters, and let $\Gamma = \Sigma \cup \{c,d\}$. Then given a finite reduction system R on Σ and two words $u,v \in \Sigma^*$, one can effectively construct a finite reduction system $R_{u,v}$ on Γ and a homomorphism $h: \Sigma^* \to \Gamma^*$ such that either

(i) $u \overset{*}{\underset{R}{\leftrightarrow}} v$ and the congruence $\overset{*}{\underset{R_{u,v}}{\longleftrightarrow}}$ is trivial,

or

(ii) $u \overset{*}{\underset{R}{\not\leftrightarrow}} v$ and h induces an embedding of M_R in $M_{R_{u,v}}$.

Here, the congruence $\overset{*}{\underset{R_{u,v}}{\longleftrightarrow}}$ is called trivial if it satisfies $x \overset{*}{\underset{R_{u,v}}{\longleftrightarrow}} y$ for all words $x,y \in \Gamma^*$.

Proof. Let R be a rewriting system on Σ, and let $u,v \in \Sigma^*$. We define a rewriting system $R_{u,v}$ by taking $R_{u,v} = R \cup \{(cud,e)\} \cup \{(acvd,cvd) \mid a \in \Gamma\}$ and a homomorphism $h: \Sigma^* \to \Gamma^*$ by $a \to a$ for all $a \in \Sigma$. Since $R \subseteq R_{u,v}$, we have $h(\ell) = \ell \overset{*}{\underset{R_{u,v}}{\longleftrightarrow}} r = h(r)$ for all $(\ell,r) \in R$, which implies that h induces a homomorphism from M_R into $M_{R_{u,v}}$. Obviously, $R_{u,v}$ and h can be constructed effectively from R and u,v. Thus, it remains to check that conditions (i) and (ii) are satisfied.

First assume that $u \overset{*}{\underset{R}{\leftrightarrow}} v$. Then for each $a \in \Gamma$, we have the following chain of congruences: $a \underset{R_{u,v}}{\longleftrightarrow} acud \overset{*}{\underset{R_{u,v}}{\longleftrightarrow}} acvd \underset{R_{u,v}}{\longleftrightarrow} cvd \overset{*}{\underset{R_{u,v}}{\longleftrightarrow}} cud \underset{R_{u,v}}{\longleftrightarrow} e$, i.e., the congruence $\overset{*}{\underset{R_{u,v}}{\longleftrightarrow}}$ is in fact trivial.

Now assume that $u \overset{*}{\underset{R}{\not\leftrightarrow}} v$. We must prove that the homomorphism $\hat{h}: M_R \to M_{R_{u,v}}$ induced by h is an embedding, i.e., it is 1-to-1. To this end it suffices to prove that for all words $x,y \in \Sigma^*$, $x \overset{*}{\underset{R_{u,v}}{\longleftrightarrow}} y$ implies $x \overset{*}{\underset{R}{\leftrightarrow}} y$, since $h(x) = x$ and $h(y) = y$.

Let $R_1 = R \cup \{(cud,e)\}$. As an intermediate step we prove the following claim.

Claim 1. For all $x,y \in \Sigma^*$, if $x \overset{*}{\underset{R_1}{\leftrightarrow}} y$, then $x \overset{*}{\underset{R}{\leftrightarrow}} y$.

Proof. If $z \in \Gamma^*$ satisfies $z \overset{*}{\underset{R_1}{\leftrightarrow}} x$, then occurrences of c and d in z function as left and right parantheses, respectively, and further, the occurrences of c and d are correctly balanced, i.e., $|z|_c = |z|_d$, and whenever $z = z_1z_2$, then $|z_1|_c \geq |z_1|_d$. Here, for $a \in \Gamma$, $|w|_a$ denotes the number of occurrences of the letter a in w.

Let $x = x_o \underset{R_1}{\leftrightarrow} x_1 \underset{R_1}{\leftrightarrow} \dots \underset{R_1}{\leftrightarrow} x_n = y$ be a derivation of minimal length of y from x in R_1. If $x_i \in \Sigma^*$ for all i, then the above derivation is actually an R-derivation implying that $x \overset{*}{\underset{R}{\leftrightarrow}} y$.

So assume that for some $i \in \{1,2,\ldots,n-1\}$, $|x_i|_c = |x_i|_d > 0$. Then there exists an index $p \in \{1,2,\ldots,n-1\}$ such that $0 = |x_0|_c \leq |x_1|_c < \ldots < |x_p|_c = |x_{p+1}|_c + 1$, i.e., x_p is the first word in the above derivation to which the rule cud $\rightarrow$ e is applied. Thus, $x_p = wcudz \underset{R_1}{\longleftrightarrow} wz = x_{p+1}$ for some words $w,z \in \Gamma^*$. Since $x = x_0 \in \Sigma^*$, and since the rules of R do not contain any occurrences of the letters c and d, the particular occurrences of c and d that are cancelled in the step from x_p to x_{p+1} must have been introduced together at some earlier step $q \in \{0,1,\ldots,p-1\}$. Hence, there exist an index $q \in \{0,1,\ldots,p-1\}$ and words $w_1,z_1 \in \Gamma^*$ such that $x_q = w_1z_1 \underset{R_1}{\longleftrightarrow} w_1cudz_1 = x_{q+1}$, $w_1 \overset{*}{\underset{R_1}{\longleftrightarrow}} w$ and $z_1 \overset{*}{\underset{R_1}{\longleftrightarrow}} z$. So the above derivation has the following form: $x = x_0 \underset{R_1}{\longleftrightarrow} \cdots \underset{R_1}{\longleftrightarrow} x_q = w_1z_1 \underset{R_1}{\longleftrightarrow} w_1cudz_1 = x_{q+1} \overset{*}{\underset{R_1}{\longleftrightarrow}} x_p = wcudz \underset{R_1}{\longleftrightarrow} wz = x_{p+1} \underset{R_1}{\longleftrightarrow} \cdots \underset{R_1}{\longleftrightarrow} x_n = y$. Since the factor cud is not touched during the derivation of x_p from x_{q+1}, we get a shorter derivation of y from x as follows: $x = x_0 \underset{R_1}{\longleftrightarrow} \cdots \underset{R_1}{\longleftrightarrow} x_q = w_1z_1 \overset{*}{\underset{R_1}{\longleftrightarrow}} wz = x_{p+1} \underset{R_1}{\longleftrightarrow} \cdots \underset{R_1}{\longleftrightarrow} x_n = y$, thus contradicting the choice of the original derivation. Therefore, we conclude that $x_i \in \Sigma^*$ for all $i \in \{0,1,\ldots,n\}$. □

Now the proof of Lemma 2.9 is completed by proving the following claim.

<u>Claim 2</u>. For all $x,y \in \Sigma^*$, if $x \overset{*}{\underset{R_{u,v}}{\longleftrightarrow}} y$, then $x \overset{*}{\underset{R_1}{\longleftrightarrow}} y$.

<u>Proof</u>. Let $x = x_0 \underset{R_{u,v}}{\longleftrightarrow} x_1 \underset{R_{u,v}}{\longleftrightarrow} \cdots \underset{R_{u,v}}{\longleftrightarrow} x_n = y$ be a derivation of minimal length of y from x in $R_{u,v}$. If this derivation consists of applications of rules of R_1 only, then it witnesses $x \overset{*}{\underset{R_1}{\longleftrightarrow}} y$, and we are done.

So assume that rules from $R_{u,v} - R_1 = \{(acvd,cvd) \mid a \in \Gamma\}$ are used as well, and let $p \in \{1,2,\ldots,n-1\}$ be the smallest index such that x_p contains cvd as a factor. Now the form of the rules in $R_{u,v} - R_1$ implies that the initial segment of length p of the above derivation is of the form $x = x_0 \underset{R_1}{\longleftrightarrow} x_1 \underset{R_1}{\longleftrightarrow} \cdots \underset{R_1}{\longleftrightarrow} x_p = wcvdz$, where $w,z \in \Gamma^*$, i.e., it is an R_1-derivation of x_p from x. According to the choice of the derivation of y from x given above, also this R_1-derivation is of minimal length. Hence, we can conclude from the proof of Claim 1, that the rule cud $\rightarrow$ e is not applied during this derivation. Since $x \in \Sigma^*$, this means that there exist an index $q \in \{0,1,2,\ldots,p-1\}$ and words

$w_1, z_1 \in \Gamma^*$ such that $x_q = w_1z_1 \xleftrightarrow[R_1]{} w_1cudz_1 = x_{q+1}$, $w_1 \xleftrightarrow[R_1]{*} w$, $u \xleftrightarrow[R_1]{*} v$, and $z_1 \xleftrightarrow[R_1]{*} z$. Since $u,v \in \Sigma^*$, Claim 1 induces $u \xleftrightarrow[R]{*} v$, thus contradicting our assumption $u \not\xleftrightarrow[R]{*} v$. Hence, we conclude that a derivation of minimal length of y from x in $R_{u,v}$ only contains applications of rules of R_1. □□

Definition 2.10. Let M_1 and M_2 be two monoids that are given through presentations $(\Sigma_1;R_1)$ and $(\Sigma_2;R_2)$, respectively, where $\Sigma_1 \cap \Sigma_2 = \emptyset$. Then the free product M_1*M_2 of the monoids M_1 and M_2 is the monoid M_R which is defined through the presentation $(\Sigma;R)$, where $\Sigma = \Sigma_1 \cup \Sigma_2$ and $R = R_1 \cup R_2$.

As can be seen easily, we have $u \xleftrightarrow[R_i]{*} v$ if and only if $u \xleftrightarrow[R]{*} v$ for all $u,v \in \Sigma_i^*$, $i = 1,2$. This yields the following fundamental property of the free product.

Theorem 2.11. Let $(\Sigma_1;R_1)$ and $(\Sigma_2;R_2)$ be two presentations such that $\Sigma_1 \cap \Sigma_2 = \emptyset$, and let $(\Sigma;R)$ denote the presentation $(\Sigma_1 \cup \Sigma_2;R_1 \cup R_2)$. Then for $i = 1,2$, the identity mapping $id_i: \Sigma_i^* \to \Sigma_i^*$ induces an embedding of the monoid M_{R_i} in the monoid $M_R = M_{R_1}*M_{R_2}$.

In particular, this induces the following corollary we will make use of in the proof of Theorem 2.7.

Corollary 2.12. Let M_1 and M_2 be two finitely presented monoids such that the word problem for the free product M_1*M_2 is decidable. Then the word problems for M_1 and M_2 are decidable.

Actually, also the reverse implication holds, i.e., two finitely presented monoids M_1 and M_2 have decidable word problems if and only if their free product M_1*M_2 has a decidable word problem. For a proof of this fact in the case of groups see e.g. [Ma-Ka-So].

Now finally we are prepared to prove Markov's result.

Proof of Theorem 2.7. Let P be a Markov property. Then there are two finitely presented monoids M_1 and M_2 such that

(1) M_1 does not have property P, and M_1 is not isomorphic to a submonoid of any finitely presented monoid having property P, and

(2) M_2 does have property P.

In addition, let M_3 be a finitely presented monoid with an undecidable word problem, and for $i = 1,2,3$, let $(\Sigma_i;R_i)$ be a finite presentation of the monoid M_i. Without loss of generality we may assume that the sets of generators Σ_1, Σ_2, and Σ_3 are pairwise disjoint.

Now the free product M_R of M_1 and M_3 is given through the presentation $(\Sigma;R)$, where $\Sigma = \Sigma_1 \cup \Sigma_3$ and $R = R_1 \cup R_3$. According to Theorem 2.11 M_1 and M_3 are embedded in M_R. Hence, M_R does not have property P due to the choice of M_1, and its word problem is undecidable due to the choice of M_3 and Corollary 2.12. On the other hand, the word problem for M_R is effectively reducible to the problem of deciding property P as we will see in the following. Hence, the latter problem is in fact undecidable.

Let c and d be two new letters, that are not contained in any of the Σ_i considered so far, and let $\Gamma = \Sigma \cup \{c,d\}$. We will describe an effective process that, given two words $u,v \in \Sigma^*$, yields a finite presentation $(\Sigma_4;R_4)$ satisfying the following equivalence:

(*) The monoid M_{R_4} presented by $(\Sigma_4;R_4)$ has property P if and only if $u \overset{*}{\underset{R}{\leftrightarrow}} v$.

So let $u,v \in \Sigma^*$. Using the construction of Lemma 2.9 we obtain a finite presentation $(\Gamma;R_{u,v})$ such that either $u \overset{*}{\underset{R}{\leftrightarrow}} v$ and the monoid $M_{R_{u,v}}$ is trivial, or $u \overset{*}{\underset{R}{\not\leftrightarrow}} v$ and the monoid M_R is embedded in $M_{R_{u,v}}$. Further, define $\Sigma_4 = \Gamma \cup \Sigma_2$ and $R_4 = R_{u,v} \cup R_2$. Then the finite presentation $(\Sigma_4;R_4)$ describes the free product M_{R_4} of the monoids $M_{R_{u,v}}$ and M_2. Obviously, the presentation $(\Sigma_4;R_4)$ can be constructed effectively from u and v, since the presentations $(\Sigma;R)$ and $(\Sigma_2;R_2)$ are given in advance.

It remains to verify that the monoid M_{R_4} presented by $(\Sigma_4;R_4)$ does indeed satisfy equivalence (*). So assume first that $u \overset{*}{\underset{R}{\leftrightarrow}} v$. Then the monoid $M_{R_{u,v}}$ is trivial, i.e., $M_{R_{u,v}} \cong \{e\}$, and hence, $M_{R_4} = M_{R_{u,v}} * M_2 \cong M_2$. Now M_2 having property P, and P being an invariant property imply that M_{R_4} does have property P. If on the other hand we have $u \overset{*}{\underset{R}{\not\leftrightarrow}} v$, Lemma 2.9 and Theorem 2.11 yield the following chain of embeddings: $M_1 \to M_R \to M_{R_{u,v}} \to M_{R_4}$. Hence, by the choice of M_1 M_{R_4} does not have property P. Thus, equivalence (*) is satisfied, i.e., the word problem for M_R is indeed effectively reducible to the problem of deciding property P. This completes the proof of Theorem 2.7. □

V.3 Automata for Reduction Systems

The decision problems concerning algebraic properties of finitely presented monoids that we stated in the previous section are all undecidable in general. However, we will prove that they can be solved when they are only considered for monoids presented by certain restricted classes of finite reduction systems. In this section we describe the classes of reduction systems that we will use, and we present some technical constructions that yield automata for recognizing certain languages that are associated with reduction systems. These constructions will be very useful when proving the decidability results mentioned above.

Recall that a finite reduction system R on Σ is called complete if it has the finite termination property and the Church-Rosser property. If R is a finite complete reduction system on Σ, then for each word $u \in \Sigma^*$, there is a unique irreducible word $\hat{u} \in \mathrm{Irr}(R)$ such that $u \xleftrightarrow[R]{*} \hat{u}$, and given u, $\hat{u}$ can be determined effectively. Since the computation of $\hat{u}$ from u is uniform in R, we conclude that the uniform word problem for finite complete reduction systems is effectively decidable:

INSTANCE: A finite complete reduction system R on Σ, and two words $u,v \in \Sigma^*$.

QUESTION: Are u and v congruent modulo R, i.e., does $u \xleftrightarrow[R]{*} v$ hold ?

In general, it is undecidable whether or not a given finite reduction system is complete. On the other hand, if only length-reducing systems are considered, then this question can be answered in polynomial time ([Ka-Kr-McN-Na]). Here a reduction system R on Σ is called length-reducing if $|\ell| > |r|$ for each rule $(\ell,r) \in R$. In what follows we will sometimes make use of reduction systems that are even more restricted. So recall that a reduction system R on Σ is called monadic if it is length-reducing and satisfies $\mathrm{right}(R) \subseteq \Sigma \cup \{e\}$, where $\mathrm{right}(R) = \{r \in \Sigma^* \mid \exists \ell \in \Sigma^*: (\ell,r) \in R\}$, and that it is called special if it is length-reducing with $\mathrm{right}(R) = \{e\}$.

Finally recall that a reduction system R on Σ is called non-redundant if for each rule $(\ell,r) \in R$, neither ℓ nor r can be reduced modulo $R-\{(\ell,r)\}$. Systems of this type are also called "reduced" by some authors. It is known that for each finite complete reduction system R on Σ, there exists a unique finite complete reduction system R' on Σ such that R' is non-redundant and equivalent to R with $\mathrm{IRR}(R') = \mathrm{Irr}(R)$. Here two reduction systems are called equivalent, if they are on the same alphabet defining the same congruence. In addition, given R the non-redundant equivalent system R' of R can be determined effectively. Thus, when dealing with finite complete reduction systems we can restrict our attention to non-redundant ones whenever this seems appropriate.

In the following we describe three constructions of automata recognizing certain languages associated with finite reduction systems. The first one has already been presented before, and therefore it is only mentioned here for reasons of completeness.

Theorem 3.1. There exists an effective construction that solves the following task:

INPUT: A finite reduction system R on Σ.

OUTPUT: A deterministic finite state acceptor recognizing the set $\mathrm{Irr}(R)$ of irreducible words modulo R.

In particular, Theorem 3.1 implies that for each finite reduction system R on Σ, the set of irreducible words is a regular language. The next construction deals with the set of descendants $L(S,R) = \{v \in \Sigma^* \mid \exists u \in S: u \overset{*}{\underset{R}{\Rightarrow}} v\}$ of a regular set $S \subseteq \Sigma^*$ with respect to a finite monadic reduction system R on Σ.

Theorem 3.2. There exists an effective construction that solves the following task:

INPUT: A finite monadic reduction system R on Σ, and a finite state acceptor A with m states recognizing the subset S of Σ^*.

OUTPUT: A finite state acceptor A^* with m states that recognizes the set $L(S,R)$.

Proof. Let R be a finite monadic reduction system on Σ. For $r \in \Sigma \cup \{e\}$, let $D(r) = \{\ell \in \Sigma^* \mid (\ell,r) \in R\}$, i.e., $D(r)$ is the set of all left-hand sides of rules of R that have right-hand side r. Further, let S be a regular subset of Σ^* that is recognized by the nondeterministic finite state acceptor $A = (Q,\Sigma,\delta,q_o,F)$, where $Q = \{q_o,q_1,\ldots,q_{m-1}\}$ is the finite set of states, $\delta: Q\times\Sigma \to P(Q)$ is the transition function, $q_o \in Q$ is the initial state, and $F \subseteq Q$ is the set of final states of A. Here $P(Q)$ denotes the power set of Q, i.e., $P(Q)$ is the set of all subsets of Q. The transition function $\delta: Q\times\Sigma \to P(Q)$ is extended to $Q\times\Sigma^* \to P(Q)$ as usual ([Ho-Ul]).

To obtain an acceptor for $L(S,R)$ we modify the acceptor A by adding certain transitions if possible. The idea of adding transitions is as follows: suppose that for some letter $a \in \Sigma$, some word $\ell \in D(a)$, and some states $q_i,q_j \in Q$, $q_j \in \delta(q_i,\ell)$. Then we must add a transition from q_i to q_j with label a, if $q_j \notin \delta(q_i,a)$. The intent of adding this transition is to capture the notion that since $\ell \underset{R}{\Rightarrow} a$, a transition from q_i to q_j with label a is equivalent to a sequence of transitions from q_i to q_j with label ℓ. Further, suppose that for some word $\ell \in D(e)$, and some states $q_i, q_j \in Q$, $q_j \in \delta(q_i, \ell)$. Then for each $a \in \Sigma$ and each $q_k \in Q$, if $q_k \in \delta(q_j, a)$, then we must add a transition from q_i to q_k with label a, if $q_k \notin \delta(q_i, a)$. The intent of adding this transition is to capture the notion that since $\ell a \underset{R}{\Rightarrow} a$, a transition from q_i to q_k with label a is equivalent to a sequence of transitions from q_i to q_k with label ℓa. In addition, if this situation occurs and q_j is a final state, then q_i also becomes a final state. This whole process must now be iterated until no further transitions can be added and no additional final states can be introduced. This means that $|\Sigma| \cdot m^2$ iterations suffice.

It is clear that this basic construction will lead to an acceptor for a subset of $L(S,R)$. On the other hand, since the process of adding transitions is iterated until no further transitions can be added, it is not difficult to show that the resulting finite state acceptor actually recognizes the set $L(S,R)$. Below a formal description of the construction outlined above is given.

```
P:    begin
         INPUT: A finite monadic reduction system R on Σ, and a finite
                state acceptor A = (Q,Σ,δ,q_o,F);
(1)      z ← 1;
(2)      while z < |Σ|·|Q|^2 do
(3)      begin for all q_i,q_j ∈ Q and all r ∈ Σ ∪ {e} do
(4)            if q_j ∈   ∪      δ(q_i,ℓ) then
                        ℓ∈D(r)
(5)            begin if r = e then
(6)                  begin for all a ∈ Σ and all q_k ∈ Q do
(7)                        if q_k ∈ δ(q_j,a) and q_k ∉ δ(q_i,a) then
(8)                           δ(q_i,a) ← δ(q_i,a) ∪ {q_k};
(9)                        if q_j ∈ F and q_i ∉ F then F ← F ∪ {q_i}
                     end
(10)                 else if q_j ∉ δ(q_i,r) then
(11)                       δ(q_i,r) ← δ(q_i,r) ∪ {q_j}
               end;
(12)           z ← z+1
         end
      end.
```

□

So for a finite monadic reduction system R on Σ and a regular set $S \subseteq \Sigma^*$, the set $L(S,R)$ is also regular, and given R and a formal specification of S, one can derive a formal specification of $L(S,R)$. In fact, this process can be performed in polynomial time as can be seen easily from its formal description.

Book, Jantzen, and Wrathall [Bo-Ja-Wr] stated Theorem 3.2 for certain infinite monadic Thue systems. The proof given here is based on a proof due to Berstel, a sketch of which appeared in Book [Bo 83]. In Book and Otto [Bo-Ot] a detailed proof for a special case is given together with some remarks on how to extend it. The algorithm presented there is also polynomial in the size of the finite state acceptor A. However, the degree of the polynomial time bound depends on the reduction system R under consideration.

Our final construction deals with the congruence class $[S]_R = \bigcup_{u \in S} [u]_R$ of a regular set $S \subseteq \Sigma^*$. It implies that this class is a deterministic context-free language, whenever S is a regular set and R is a finite monadic complete reduction system, a result that also stems from Book, Jantzen and Wrathall [Bo-Ja-Wr]. However, the proof given below is taken from [Bo 84].

Theorem 3.3. There exists an effective construction that solves the following task:

INPUT: A finite monadic complete reduction system R on Σ, and a finite state acceptor A recognizing the subset S of Σ^*.

OUTPUT: A deterministic pushdown automaton B that recognizes the congruence class $[S]_R$.

Proof. Given a finite monadic complete reduction system R on Σ and a finite state acceptor A recognizing the subset S of Σ^*, we can construct a finite state acceptor A^* for recognizing the set $L(S,R)$ by applying the construction of Theorem 3.2. Using the construction of Theorem 3.1 we also get a finite state acceptor C that recognizes the set $Irr(R)$ of irreducible words modulo R. By combining A^* and C using well-known techniques from automata theory we now obtain a deterministic finite state acceptor $A_1 = (Q_1, \Sigma, \delta_1, q_0, F_1)$ recognizing the set $S_1 = L(S,R) \cap Irr(R)$.

Since the reduction system R is complete, a word $u \in \Sigma^*$ is in $[S]_R$ if and only if its irreducible descendant modulo R is in S_1. Thus, for solving the membership problem for $[S]_R$ we can proceed as follows: Given a word $u \in \Sigma^*$, u is reduced to determine its irreducible descendant w, and then $w \in S_1$ is tested using the deterministic finite state acceptor A_1. It remains to prove that this task can be performed by a deterministic pushdown automaton B that is effectively constructible from R and A_1.

Let # be a letter that is not already contained in Σ. Then the set $Q_1 \times (\Sigma \cup \{\#\})$ is to be the set of pushdown symbols of B, where $(q_o,\#)$ serves as the initial symbol on the pushdown store. The transitions of B are organized in such a manner that whenever the pushdown store contains $(q_o,\#),(q_1,a_1),(q_2,a_2),\ldots,(q_m,a_m)$, where $(q_o,\#)$ is on the bottom, and $q_1,\ldots,q_m \in Q_1$, $a_1,\ldots,a_m \in \Sigma$, then we have $\delta_1(q_i,a_{i+1}) = q_{i+1}$ for all i, $0 \leq i < m$. To facilitate this behavior the first component of the top-most symbol on the pushdown store at each given moment is also part of the actual state of B, i.e., if the top-most symbol of the pushdown store is (q,a), then q is also stored as part of B's actual state. Now we say that a letter $a' \in \Sigma$ is pushed onto the pushdown store to mean that the symbol $(\delta_1(q,a'),a')$ is pushed onto the store, where $q \in Q_1$ is the state symbol of A_1 that is part of B's actual state at this very moment. Using this convention the behavior of B can be described by the following two procedures READ and SEARCH. Initially a READ operation is attempted.

READ: If the input tape is non-empty, read an input symbol, push it onto the pushdown store, and go to SEARCH; otherwise, accept if and only if the state symbol of A_1 that is part of B's actual state belongs to F_1.

SEARCH: Pop from the pushdown store the longest string ℓ (if any) that occurs as left-hand side of a rule (ℓ,r) in R. If no such ℓ is detected, then restore the pushdown store to its previous condition and go to READ. If such an ℓ is detected, then push the corresponding r onto the pushdown store (we assume an ordering of rules with ℓ on the left-hand side and choose the first such r). Go to SEARCH.

Obviously, the above construction yields a deterministic pushdown automaton B. Now given input $u \in \Sigma^*$, B finally halts with empty input tape, its pushdown store containing $(q_o,\#),(q_1,a_1),\ldots,(q_m,a_m)$, where $w = a_1a_2\ldots a_m$ is an irreducible word modulo R. Actually, w is determined from u by a left-most reduction, and since R is complete, w is the unique irreducible descendant of u modulo R. Hence, $u \in [S]_R$ if and only if $w \in S_1 = L(S,R) \cap \mathrm{Irr}(R)$, which according to the construction of B holds if and only if $q_m \in F_1$, since $q_m = \delta_1(q_o,w)$. Hence, B accepts on input u if and only if $u \in [S]_R$. This completes the proof of Theorem 3.3. □

V.4. Deciding Algebraic Properties of Monoids Presented by Finite Complete Reduction Systems

As we saw all the problems listed at the beginning of Section 2 are undecidable in general. On the other hand, when a monoid M is given through a presentation $(\Sigma;R)$, where R is a finite complete reduction system, then for each element $m \in M$, there is a unique normal form $u_m \in \mathrm{Irr}(R)$. How much does this property help in solving the decision problems mentioned before ? Here we will see that problems 1 to 3 and 5 to 6 become decidable in this situation, while in the next section the remaining problems will be shown to become decidable when being restricted to monoids presented by finite monadic complete reduction systems.

Let R be a finite complete reduction system on Σ. Then obviously the following three statements about the monoid M_R presented by $(\Sigma;R)$ hold:

(1) M_R is trivial if and only if $a \overset{*}{\underset{R}{\leftrightarrow}} e$ for all $a \in \Sigma$.
(2) M_R is finite if and only if the set $\mathrm{Irr}(R)$ is finite.
(3) M_R is commutative if and only if $ab \overset{*}{\underset{R}{\leftrightarrow}} ba$ for all $a,b \in \Sigma$.

Given R, a deterministic finite state acceptor recognizing the set $\mathrm{Irr}(R)$ can be constructed effectively, from which the cardinality of this set can easily be determined. In addition, the uniform word problem for finite complete reduction systems is effectively decidable. Together these observations induce the following result.

Theorem 4.1. The following problems are decidable:

INSTANCE: A finite presentation $(\Sigma;R)$, where R is a complete reduction system on Σ.
1. QUESTION: Is the monoid M_R trivial ?
2. QUESTION: Is the monoid M_R finite ?
3. QUESTION: Is the monoid M_R commutative ?

In contrast to the results just given Narendran and O'Dunlaing have observed that the problem of deciding whether the monoid M_R given through a finite presentation $(\Sigma;R)$ is cancellative or not remains undecidable, even if it is restricted to presentations involving complete reduction systems that contain length-reducing rules only [Na-O'Dun].

Now we want to turn to problems 5 and 6. Is it decidable whether or not a monoid given through a finite complete reduction system is a free monoid, and is it decidable whether or not it is a group ? So far only partial results concerning these problems were known. In [Ot-1] it is shown that the former problem can be solved in linear space for all presentations involving finite length-reducing complete reduction systems, and Book [Bo 82] proves that the latter problem is solvable in polynomial time for all presentations involving finite monadic complete reduction systems, and that it is solvable in real time for all presentations involving finite reduction systems that are special and complete. In the following we want to prove that these problems become decidable when they are restricted to monoids given through finite complete reduction systems. Recall that a finitely presented monoid M is <u>free</u> if and only if it has a presentation of the form $(\Gamma;\emptyset)$ for some finite alphabet Γ, i.e., if and only if it is isomorphic to the monoid Γ^*.

So let R be a fixed finite complete reduction system on a finite alphabet Σ. We are interested in finding out whether or not the monoid M defined by the presentation $(\Sigma;R)$ is free. For doing so we first transform the given presentation into a certain normal form. Using the properties of this normal form we can then characterize a set of free generators of the monoid M, provided M is indeed a free monoid. Finally, we show how this characterization can be turned into an algorithm for testing whether or not M is actually free.

Since for each finite complete reduction system, there is an equivalent one that is finite, complete, and non-redundant and that can be determined effectively from the former, the first step in our normalization of the given presentation $(\Sigma;R)$ consists in computing a finite complete non-redundant reduction system R_o on Σ such that R and R_o are equivalent. Then the ordered pair $(\Sigma;R_o)$ is another finite presentation of the monoid M. Further, R_o being complete implies that the empty word e does not occur as left-hand side of a rule of R_o, i.e., $\text{left}(R_o) \subseteq \Sigma^+$.

Now assume that the reduction system R_o contains a rule of the form (a,w) for some $a \in \Sigma$. Since R_o is non-redundant and complete, this means that the letter a does neither occur in w nor in any other rule of R_o. Hence, application of an elementary Tietze transformation of type 4 results in the presentation $(\Sigma-\{a\};R_o-\{(a,w)\})$. Obviously, $R_o-\{(a,w)\}$ is a finite complete non-redundant reduction system on $\Sigma-\{a\}$, and by Lemma 1.3 the presentation $(\Sigma-\{a\};R_o-\{(a,w)\})$ still defines the monoid M. Iterating this process we finally obtain a finite presentation $(\Sigma_1;R_1)$ of M such that R_1 is a non-redundant complete reduction system satisfying $\text{left}(R_1) \cap \Sigma_1 = \emptyset$. Thus, the left-hand side of each rule of R_1 is a word of length at least 2. The presentation

$(\Sigma_1;R_1)$ is a normal form of the presentation $(\Sigma;R)$.

Lemma 4.2. If the monoid M_R given by a finite presentation of the form $(\Sigma;R)$ is free, then there exists a subset Σ_o of Σ that freely generates this monoid.

Proof. Let $\Sigma = \{a_1,a_2,\ldots,a_n\}$, and let R be a finite reduction system on Σ such that the monoid M_R presented by $(\Sigma;R)$ is free. Then it is free of rank m for some $m < n$. Hence, there exists an alphabet $\Gamma = \{b_1,b_2,\ldots,b_m\}$ of cardinality m such that $M_R = \Sigma^*/\overset{*}{\underset{R}{\longleftrightarrow}} \cong \Gamma^*$. So for each $a_i \in \Sigma$, there exists a word $u_i \in \Gamma^*$ such that a_i and u_i represent the same element of M_R. Analogously, for each $b_j \in \Gamma$, there exists a word $v_j \in \Sigma^*$ such that b_j and v_j represent the same element of M_R. Since $M_R \cong \Gamma^*$, no $b_j \in \Gamma$ represents the identity of M_R, and so $v_j \neq e$, $j = 1,2,\ldots,m$. Further, the words $v_j \in \Sigma^*$ can be chosen in such a way that no v_j contains an occurrence of a letter $a_i \in \Sigma$ with $a_i \overset{*}{\underset{R}{\longleftrightarrow}} e$. On the other hand, $u_i = e$ if and only if $a_i \overset{*}{\underset{R}{\longleftrightarrow}} e$, $i = 1,2,\ldots,n$.

Let $b_j \in \Gamma$, and assume that $v_j = a_{i_1}a_{i_2}\ldots a_{i_k}$ for some letters $a_{i_1},a_{i_2},\ldots,a_{i_k} \in \Sigma$. For each $\ell \in \{1,2,\ldots,k\}$, u_{i_ℓ} represents the same element of M_R as a_{i_ℓ}. Hence, b_j and the word $u_{i_1}u_{i_2}\ldots u_{i_k} \in \Gamma^*$ represent the same element of M_R. Since $M_R \cong \Gamma^*$, this implies that $b_j = u_{i_1}u_{i_2}\ldots u_{i_k}$. By the choice of v_j, we have $u_{i_\ell} \neq e$ for all ℓ implying that $k = 1$, i.e., $v_j \in \Sigma$.

Take $\Sigma_o = \{a \in \Sigma \mid \exists b_j \in \Gamma: a = v_j\}$. Then $\Sigma_o = \{v_j \mid j = 1,2,\ldots,m\}$ is a subset of Σ that freely generates M_R. □

Assume that the monoid M is free. Then there exists a subset Σ_o of Σ_1 that freely generates M. Hence, for each letter $a \in \Sigma_1$, there is a unique word $u_a \in \Sigma_o^*$ such that $a \overset{*}{\underset{R_1}{\longleftrightarrow}} u_a$, i.e., a and u_a define the same element of M. Define a homomorphism $\varphi\colon \Sigma_1^* \to \Sigma_o^*$ by taking $\varphi(a) = u_a$ for all $a \in \Sigma_1$. Then we have the following.

Lemma 4.3. For all $u,v \in \Sigma_1^*$, $u \overset{*}{\underset{R_1}{\longleftrightarrow}} v$ if and only if $\varphi(u) = \varphi(v)$.

Proof. Since $\varphi(w) \overset{*}{\underset{R_1}{\longleftrightarrow}} w$ for all $w \in \Sigma_1^*$, $u \overset{*}{\underset{R_1}{\longleftrightarrow}} v$ if and only if $\varphi(u) \overset{*}{\underset{R_1}{\longleftrightarrow}} \varphi(v)$. But $\varphi(u),\varphi(v) \in \Sigma_o^*$, and Σ_o freely generates M implying that $\varphi(u) \overset{*}{\underset{R_1}{\longleftrightarrow}} \varphi(v)$ if and only if $\varphi(u) = \varphi(v)$. □

Since $|\ell| > 2$ for each rule $(\ell,r) \in R_1$ according to the normalization we performed, we have $\Sigma_1 \subseteq \mathrm{Irr}(R_1)$. This means in particular, that $u_a \neq e$ for each letter $a \in \Sigma_1$. Hence, we can conclude the following from Lemma 4.3.

Corollary 4.4. For each letter $a \in \Sigma_1$ and each word $u \in \Sigma_1^*$, if $u \overset{*}{\underset{R_1}{\leftrightarrow}} a$, then either $u = a$ or $|u|_a = 0$.

The following two lemmas will help us to derive a characterization of the set Σ_o that freely generates the monoid M.

Lemma 4.5. $\mathrm{right}(R_1) \subseteq \Sigma_1^+$, i.e., R_1 does not contain any rule of the form (ℓ,e).

Proof. Assume to the contrary that $(\ell,e) \in R_1$ for some word $\ell \in \Sigma_1^*$. Then $\ell \overset{*}{\underset{R_1}{\leftrightarrow}} e$, and hence $\varphi(\ell) = \varphi(e) = e$ by Lemma 4.3. On the other hand, we have $|\varphi(\ell)| \geq |\ell| \geq 2$ contradicting the equality just stated. □

Finally, let $\Sigma_2 = \Sigma_1 - \Sigma_o$. Then this subset of Σ_1 can be described as follows.

Lemma 4.6. $\Sigma_2 = \mathrm{right}(R_1) \cap \Sigma_1$, i.e., for each $a \in \Sigma_2$, there exists a word $\ell \in \Sigma_1^*$ such that $(\ell,a) \in R_1$, and furthermore, these are the only rules of R_1 with right-hand sides of length 1.

Proof. Since $\Sigma_1 \subseteq \mathrm{Irr}(R_1)$, we can conclude that $u_a \overset{*}{\underset{R_1}{\to}} a$ for each letter $a \in \Sigma_1$. For $a \in \Sigma_2$, the fact that $u_a \in \Sigma_o^+$ then implies that $u_a \overset{+}{\underset{R_1}{\to}} a$, which yields $a \in \mathrm{right}(R_1)$ by Lemma 4.5. Thus, $\Sigma_2 \subseteq \mathrm{right}(R_1) \cap \Sigma_1$.

Now assume that $(\ell,b) \in R_1$ for some letter $b \in \Sigma_o$. Then $|\ell| \geq 2$, and since $u_a \in \Sigma_o^+$ for each letter $a \in \Sigma_1$, this gives $|\varphi(\ell)| \geq |\ell| \geq 2$, while $|\varphi(b)| = |b| = 1$, thus contradicting Lemma 4.3. Hence $\Sigma_2 = \mathrm{right}(R_1) \cap \Sigma_1$. □

Thus, $\Sigma_o = \Sigma_1 - (\mathrm{right}(R_1) \cap \Sigma_1)$, i.e., Σ_o can easily be obtained from R_1. Hence, the monoid M is free if and only if it is freely generated by the set $\Sigma_o = \Sigma_1 - (\mathrm{right}(R_1) \cap \Sigma_1)$. It remains to show that it is effectively decidable whether or not the set Σ_o freely generates the monoid M. To this end observe that Lemma 4.6 does not only characterize the letters that we must try to get rid off, but together with Corollary 4.4 it also shows that these letters can be eliminated by elementary Tietze transformations of type 4 provided the monoid M is free.

So we are going to present an algorithm that performs the following task: If the monoid M is free, then this algorithm will transform the presentation $(\Sigma_1;R_1)$ into a presentation that clearly indicates the fact that M is a free monoid. On the other hand, if M is not free, then this algorithm will realize this fact at some stage of its computation, and it will then reject the input.

<u>Algorithm 4.7</u>.

INPUT: A finite alphabet Σ_1, and a finite complete non-redundant reduction system R_1 on Σ_1 such that $\Sigma_1 \subseteq \mathrm{Irr}(R_1)$;

(1) <u>begin</u> <u>if</u> $e \in \mathrm{right}(R_1)$ <u>then</u> reject;

(2) $\Sigma_2 \leftarrow \mathrm{right}(R_1) \cap \Sigma_1$;

(3) <u>while</u> $\Sigma_2 \neq \emptyset$ <u>do</u>

(4) <u>begin</u> Choose a letter $a \in \Sigma_2$ together with a rule $(\ell,a) \in R_1$;

(5) <u>if</u> ℓ contains an occurrence of the letter a <u>then</u> <u>reject</u>;

(6) $\Sigma_2 \leftarrow \Sigma_2 - \{a\}$;

(7) $\Sigma_1 \leftarrow \Sigma_1 - \{a\}$;

(8) $R_1 \leftarrow R_1 - \{(\ell,a)\}$;

(9) Substitute each occurrence of a in each rule of R_1 by the word ℓ

<u>end</u>;

(10) <u>if</u> $R_1 \subseteq \{(w,w) \mid w \in \Sigma^*\}$ <u>then</u> <u>accept</u> <u>else</u> <u>reject</u>

<u>end</u>.

<u>Lemma 4.8</u>. Let R_1 be a finite complete non-redundant reduction system on Σ_1 such that $\Sigma_1 \subseteq \mathrm{Irr}(R_1)$. Then algorithm 4.7 accepts on input $(\Sigma_1;R_1)$ if and only if the monoid M presented by $(\Sigma_1;R_1)$ is a free monoid.

<u>Proof</u>. If M is a free monoid, then there is a subset Σ_o of Σ_1 that freely generates M, and by Lemma 4.5 R_1 contains no rule with right-hand side e, i.e., $e \notin \mathrm{right}(R_1)$. Hence, algorithm 4.7 does not reject in line (1). Let $\Sigma_2 := \Sigma_1 - \Sigma_o$. By Lemma 4.6 we have $\Sigma_2 = \mathrm{right}(R_1) \cap \Sigma_1$. For $a \in \Sigma_2$, we thus have at least one rule with right-hand side a. So let $\ell \in \Sigma_1^*$ such that $(\ell,a) \in R_1$. Then $|\ell| > 2$ implying that $|\ell|_a = 0$ by Corollary 4.4. Hence, algorithm 4.7 does not reject in line (5). In lines (6) to (9) an elementary Tietze transformation of type 4 is performed that results in deleting the letter a and the rule (ℓ,a) from the presentation $(\Sigma_1;R_1)$. Also, a is deleted from the set Σ_2 giving the set Σ_2'.

Let $(\Sigma';R')$ denote the presentation of M this Tietze transformation yields. Then for each $b \in \Sigma_2'$, R' contains at least one rule of the form (v,b) with $v \in \Sigma'^*$. Now $v \in \Sigma'^* = (\Sigma_o \cup \Sigma_2')^* \subseteq \Sigma_1^*$ and $v \overset{*}{\underset{R_1}{\leftrightarrow}} b$, which imply that $|v|_b = 0$ by Corollary 4.4. Thus, the loop (3) - (9) is performed until $\Sigma_2 = \emptyset$ is reached without rejecting in line (5).

When this loop is left, the presentation of M we have at this moment is of the form $(\Sigma_o;R')$ for some reduction system $R' \subseteq \Sigma_o^* \times \Sigma_o^*$.

But M is freely generated by Σ_o, which implies that $R' \subseteq \{(w,w) \mid w \in \Sigma_o^*\}$, i.e., algorithm 4.7 accepts.

Now assume to the contrary that algorithm 4.7 accepts on input $(\Sigma_1;R_1)$. Then the presentation $(\Sigma_1;R_1)$ is transformed into a presentation of the form $(\Sigma_o;R')$ for some subset Σ_o of Σ_1 and some reduction system $R' \subseteq \{(w,w) \mid w \in \Sigma_o^*\}$. But during this process only Tietze transformations are applied to presentations of M starting with $(\Sigma_1;R_1)$. Hence, $(\Sigma_o;R')$ is itself a presentation of M, and so M is a free monoid that is freely generated by Σ_o. □

Hence, we have shown the following result.

Theorem 4.9. The following problem is decidable:

INSTANCE: A finite presentation $(\Sigma;R)$, where R is a complete reduction system on Σ.

QUESTION: Is the monoid M_R given through this presentation a free monoid ?

Before turning to the next decision problem we give an example of a non-trivial presentation of a free monoid.

Example 4.10. Let $\Sigma = \{a,b,c,d_1,d_2,d_3,f\}$ and $R = \{(abc,d_1),(ad_1^2c,d_2),(ad_2^2c,d_3),(d_2cb,f),(d_3b,ad_2f)\}$. Then R is a finite non-redundant reduction system on Σ, and it can be checked easily that R is noetherian and confluent.

Now $\Sigma_2 := \text{right}(R) \cap \Sigma = \{d_1,d_2,d_3,f\}$, and $\Sigma_o := \Sigma - \Sigma_2 = \{a,b,c\}$. On input $(\Sigma;R)$ algorithm 4.7 yields the presentation $(\Sigma_o;\{(a^3bcabc^2a^2bcabc^3b,a^3bcabc^2a^2bcabc^3b)\})$, thus proving that the monoid M_R presented by $(\Sigma;R)$ is a free monoid. □

Observe that we do not only decide whether or not a monoid given by a presentation $(\Sigma;R)$, where R is a finite complete reduction system on Σ, is free, but in the affirmative we also determine the rank of this free monoid by displaying a set of free generators.

Finally, we want to show that the problem of deciding of whether or not a monoid given through a finite presentation is a group also becomes decidable when it is restricted to presentations involving finite complete reduction systems. So let R be a finite complete reduction system on Σ. Then the monoid M given through the presentation $(\Sigma;R)$ is a group if and only if, for each word $w \in \Sigma^*$, there exists a word $w' \in \Sigma^*$ such that $ww' \xleftrightarrow[R]{*} e$. Obviously, this is equivalent to saying that for each letter $a \in \Sigma$, there exists a word $u_a \in \Sigma^*$ such that $au_a \overset{*}{\underset{R}{\longleftrightarrow}} e$. In the following we will show how to effectively determine reasonable candidates for these words. Using the fact that R is a complete reduction system, we can then easily check whether these candidates actually satisfy the above congruences.

As a first step we define a sequence $\Sigma_1, \Sigma_2, \ldots, \Sigma_i, \ldots$ of subsets of Σ inductively by taking

$$\Sigma_1 := \{a \in \Sigma \mid \exists u,v \in \Sigma^*: (uav,e) \in R\}$$

and

$$\Sigma_{i+1} := \{a \in \Sigma-\Sigma_i \mid \exists u,v \in \Sigma^* \; \exists w \in \Sigma_i^*: (uav,w) \in R \text{ or } (w,uav) \in R\} \cup \Sigma_i.$$

Then we have $\Sigma_1 \subseteq \Sigma_2 \subseteq \ldots \subseteq \Sigma_i \subseteq \Sigma_{i+1} \subseteq \ldots \subseteq \bigcup_{j\geq 1} \Sigma_j \subseteq \Sigma$. Obviously, if $\Sigma_i = \Sigma_{i+1}$ for some $i \geq 1$, then $\Sigma_i = \bigcup_{j\geq 1} \Sigma_j$. Since Σ is finite, we conclude that the above chain of inclusions contains a finite number of different sets only. Thus, it suffices to construct $\Sigma_1, \Sigma_2, \ldots, \Sigma_n$, where n is the cardinality of Σ, and this can be done effectively.

Now for each letter $a \in \Sigma_n$, we determine a word $u_a \in \Sigma^*$ as follows. If $a \in \Sigma_1$, then $(uav,e) \in R$ for some words $u,v \in \Sigma^*$. We choose one such rule, and take $u_a := vu$. If $a \in \Sigma_{i+1}-\Sigma_i$ for some $i \geq 1$, then $(uav,w) \in R$ or $(w,uav) \in R$ for some words $u,v \in \Sigma^*$ and $w \in \Sigma_i^*$. Again we choose one such rule. Let $w = a_{i_1} a_{i_2} \ldots a_{i_m} \in \Sigma_i^*$. Then words $u_{a_{i_1}}, u_{a_{i_2}}, \ldots, u_{a_{i_m}} \in \Sigma^*$ have already been chosen. Now we take $u_a := vu_{a_{i_m}} \ldots u_{a_{i_2}} u_{a_{i_1}} u$. Notice that in this way, for each letter $a \in \Sigma_n$, a corresponding word u_a is determined effectively.

Lemma 4.11. The monoid M given through the presentation $(\Sigma;R)$ is a group if and only if the following two conditions are satisfied:

(i) $\Sigma = \Sigma_n$, where n denotes the cardinality of Σ;

(ii) $au_a \overset{*}{\underset{R}{\longleftrightarrow}} e$ for all $a \in \Sigma_n$.

Proof. If conditions (i) and (ii) are satisfied, then M is obviously a group. Thus, it remains to prove the converse implication. So assume that M is a group.

Let $a \in \Sigma$. Since M is a group, there exists a word $w \in \Sigma^*$ such that $aw \overset{*}{\underset{R}{\longleftrightarrow}} e$. R is complete, and so $aw \overset{*}{\underset{R}{\to}} e$, i.e., there exists a sequence of reductions of the form $aw = v_m \underset{R}{\to} v_{m-1} \underset{R}{\to} \cdots \underset{R}{\to} v_1 \underset{R}{\to} e$. Now $(v_1,e) \in R$ yields that $v_1 \in \Sigma_1^*$. For each $i \geq 1$, $v_{i+1} \underset{R}{\to} v_i$ implying that $v_{i+1} = x\ell y$ and $v_i = xry$ for some words $x,y \in \Sigma^*$ and a rule (ℓ,r) of R. If $v_i \in \Sigma_i^*$, we have $r \in \Sigma_i^*$, which in turn gives $\ell \in \Sigma_{i+1}^*$. Hence, if $v_i \in \Sigma_i^*$, then $v_{i+1} \in \Sigma_{i+1}^*$. By induction this implies $aw = v_m \in \Sigma_m^*$. But $\Sigma_1 \subseteq \Sigma_2 \subseteq \ldots \subseteq \Sigma_n = \bigcup_{j\geq 1}\Sigma_j \subseteq \Sigma$, and so we actually see that $aw \in \Sigma_n^*$, i.e., $a \in \Sigma_n$. Thus, condition (i) is satisfied.

In addition, we must show that condition (ii) also holds. To do so, we prove by induction on i, that for each $i \geq 1$ and each $a \in \Sigma_i$, $au_a \overset{*}{\underset{R}{\leftrightarrow}} e$ holds. Let $a \in \Sigma_1$. Then $u_a = vu$, where $u,v \in \Sigma^*$ are words such that (uav,e) is a rule of R. This gives $uav \overset{*}{\underset{R}{\leftrightarrow}} e$, which in turn implies $au_a = avu \overset{*}{\underset{R}{\leftrightarrow}} e$, since M is a group. If $a \in \Sigma_{i+1}-\Sigma_i$, then $u_a = vxu$, where $u,v \in \Sigma^*$ and $x = u_{a_{i_m}} \ldots u_{a_{i_2}} u_{a_{i_1}}$ with $a_{i_1}, a_{i_2}, \ldots, a_{i_m} \in \Sigma_i$ such that $(uav, a_{i_1}a_{i_2}\ldots a_{i_m}) \in R$ or $(a_{i_1}a_{i_2}\ldots a_{i_m}, uav) \in R$. Hence, $uavx \overset{*}{\underset{R}{\leftrightarrow}} a_{i_1}a_{i_2}\ldots a_{i_m} x = a_{i_1}a_{i_2}\ldots a_{i_m} u_{a_{i_m}} \ldots u_{a_{i_1}} \overset{*}{\underset{R}{\leftrightarrow}} e$ by induction hypothesis. Since M is a group, this implies $au_a = avxu \overset{*}{\underset{R}{\leftrightarrow}} e$, thus completing the proof of Lemma 4.11. □

Since the set Σ_n and the words u_a $(a \in \Sigma_n)$ can be computed effectively, and since the uniform word problem for finite complete reduction systems is decidable, we have thus derived the following result.

<u>Theorem 4.12</u>. The following problem is decidable:

INSTANCE: A finite presentation $(\Sigma;R)$, where R is a complete reduction system on Σ.

QUESTION: Is the monoid M_R given through this presentation a group ?

Given an alphabet Σ of cardinality $n \in \mathbb{N}$ and a finite, complete reduction system R on Σ, the set Σ_n can be determined in polynomial time. As can be seen easily by induction on i, $|u_a| < \mu^i$ for all $a \in \Sigma_i$, where $\mu = \max\{|\ell|,|r| \mid (\ell,r) \in R\}+1$. So it may take up to exponential time to determine the words u_a $(a \in \Sigma_n)$. Finally, the complexity of checking of whether or not $au_a \overset{*}{\underset{R}{\leftrightarrow}} e$ holds depends on the complexity of the word problem for R which may be arbitrarily high as shown by Bauer and Otto [Ba-Ot].

<u>Example 4.13</u>. Let $n \geq 1$ be an integer, and let $\Sigma = \{a_1,b_1,c_1,a_2,b_2,c_2,\ldots,a_n,b_n,c_n\}$ be an alphabet of cardinality $|\Sigma| = 3n$. Further, let $R = \{(a_1b_1c_1,e),(a_2b_2c_2,a_1c_1),$

$(a_3b_3c_3,a_2c_2),\ldots,(a_nb_nc_n,a_{n-1}c_{n-1})\}$. Then R is a finite complete reduction system on Σ. In addition, R is non-redundant.

For Σ we get the following sequence $\Sigma_1,\Sigma_2,\ldots,\Sigma_i,\ldots$ of subsets of Σ: $\Sigma_1 = \{a_1,b_1,c_1\}$, $\Sigma_2 = \{a_2,b_2,c_2\} \cup \Sigma_1,\ldots,\Sigma_n = \{a_n,b_n,c_n\} \cup \Sigma_{n-1} = \Sigma$. In particular, condition (i) of Lemma 4.11 is satisfied. Further, we get the following words u_a for $a \in \Sigma_n$:

$u_{a_1} = b_1c_1,\ u_{b_1} = c_1a_1,\ u_{c_1} = a_1b_1,$

$u_{a_2} = b_2c_2u_{c_1}u_{a_1},\ u_{b_2} = c_2u_{c_1}u_{a_1}a_2,\ u_{c_2} = u_{c_1}u_{a_1}a_2b_2,$

...

$u_{a_n} = b_nc_nu_{c_{n-1}}u_{a_{n-1}},\ u_{b_n} = c_nu_{c_{n-1}}u_{a_{n-1}}a_n,\ u_{c_n} = u_{c_{n-1}}u_{a_{n-1}}a_nb_n.$

Hence, for each $i \in \{1,2,\ldots,n\}$, $|u_{a_i}| = |u_{b_i}| = |u_{c_i}| = 2^{i+1}-2$, i.e., some of the candidates for the inverses of letters of Σ have exponential length. However it can be checked very easily that condition (ii) of Lemma 4.11 is not satisfied, since $b_1u_{b_1} = b_1c_1a_1 \not\overset{*}{\underset{R}{\leftrightarrow}} e$. Thus, the monoid M_R presented by $(\Sigma;R)$ is not a group. □

We have seen that both problems 5 and 6 become decidable, when they are restricted to finite presentations $(\Sigma;R)$ involving complete reduction systems. However, for constructing the candidates u_a for the inverses of the letters a ($a \in \Sigma_n$), the property of R of being complete was not used at all. In fact, Lemma 4.11 holds for all finite presentations. Thus, in order to be able to check whether or not the monoid M_R presented by $(\Sigma;R)$ is a group, it is sufficient that the following <u>restricted version of the word problem</u> for $(\Sigma;R)$ is decidable:

INSTANCE: A word $w \in \Sigma^*$.
QUESTION: Does $w \overset{*}{\underset{R}{\leftrightarrow}} e$ hold ?

Also the result concerning problem 5 can be carried over to finite presentations, for which the restricted version of the word problem stated above is decidable. Thus, the results obtained can be viewed as special cases of the following general result.

<u>Theorem 4.14</u> ([Ot 86]) . Let C be a class of finite presentations such that there exists an algorithm $A(C)$ satisfying

$$A(C)((\Sigma;R),w) = \begin{cases} 1 & \text{if } w \overset{*}{\underset{R}{\leftrightarrow}} e \\ 0 & \text{if } w \not\overset{*}{\underset{R}{\leftrightarrow}} e \end{cases}$$

for all presentations $(\Sigma;R)$ from C and all words $w \in \Sigma^*$. Then it is decidable whether or not a given presentation $(\Sigma;R)$ from C defines a free monoid or a group.

V.5. Deciding Algebraic Properties of Monoids Presented by Finite Monadic Complete Reduction Systems

In [Bo 83] Book has established a syntactic class of formulae, called linear sentences over Σ, such that each reduction system R on Σ induces an interpretation of these sentences. Hence, these sentences can be used to describe certain properties of the congruence generated by R. The main result of [Bo 83] states that it is decidable whether or not a given linear sentence is true under the interpretation induced by a given finite monadic complete reduction system. Thus, for a finite monadic complete reduction system all those decision problems can be solved by using the technique of linear sentences that are expressible by these sentences. Problems of this kind are the left and right divisibility problems, the power problem, the membership problem for a finitely generated submonoid, the independent set problem for finite sets, etc. Unfortunately it is not known whether the properties of cancellativity, idempotency, and torsion-freeness can be expressed by linear sentences. Hence, in order to solve our remaining five decision problems at least for monoids presented by finite monadic complete reduction systems, we must devise different techniques.

Definition 5.1. Let M be a monoid presented by $(\Sigma;R)$. M is called left-cancellative if, for all $x,y,z \in \Sigma^*$, $xy \overset{*}{\underset{R}{\leftrightarrow}} xz$ implies $y \overset{*}{\underset{R}{\leftrightarrow}} z$. It is called right-cancellative if, for all $x,y,z \in \Sigma^*$, $xz \overset{*}{\underset{R}{\leftrightarrow}} yz$ implies $x \overset{*}{\underset{R}{\leftrightarrow}} y$. Finally, M is called cancellative, if it is both left-cancellative and right-cancellative.

In what follows we will be dealing only with the property of left-cancellativity, but by symmetry according results can easily be obtained for the property of right-cancellativity. The results presented are taken from Narendran and O'Dunlaing [Na-O'Dun].

Lemma 5.2. Let M be a monoid presented by $(\Sigma;R)$. Then the following two statements are equivalent:

(i) M is left-cancellative;

(ii) for all letters $a \in \Sigma$ and all words $x,y \in \Sigma^*$, $ax \overset{*}{\underset{R}{\leftrightarrow}} ay$ implies $x \overset{*}{\underset{R}{\leftrightarrow}} y$.

Proof. Obviously, (i) implies (ii). To prove the reverse implication assume that the monoid M is not left-cancellative, and let $x \in \Sigma^+$ be a word of minimal length such that there exist $y,z \in \Sigma^*$ satisfying $xy \overset{*}{\underset{R}{\leftrightarrow}} xz$ and $y \overset{*}{\underset{R}{\not\leftrightarrow}} z$. If $|x| > 1$, we can write x as $x = au$ for some letter $a \in \Sigma$ and some word $u \in \Sigma^+$. Now either $uy \overset{*}{\underset{R}{\leftrightarrow}} uz$ or $uy \overset{*}{\underset{R}{\not\leftrightarrow}} uz$. In the first case we have $uy \overset{*}{\underset{R}{\leftrightarrow}} uz$ and $y \overset{*}{\underset{R}{\not\leftrightarrow}} z$ contradicting the choice of x, and in the second case we have $a(uy) = xy \overset{*}{\underset{R}{\leftrightarrow}} xz = a(uz)$ and $uy \overset{*}{\underset{R}{\not\leftrightarrow}} uz$ again contradicting the choice of x. Thus, $|x| = 1$, i.e., (ii) is not satisfied either. □

From Lemma 5.2 we can immediately conclude the following.

Corollary 5.3. Let R be a complete reduction system on Σ. Then the monoid M presented by $(\Sigma;R)$ is left-cancellative if and only if the following condition holds:
For all letters $a \in \Sigma$ and all words $x,y \in \mathrm{Irr}(R)$, $ax \overset{*}{\underset{R}{\leftrightarrow}} ay$ implies $x = y$.

In particular, this means that whenever a non-redundant complete reduction system R presents a left-cancellative monoid, then R cannot contain a rule of the form (ax,ay) where $a \in \Sigma$.

Lemma 5.4. Let R be a length-reducing complete reduction system on Σ. If the monoid M presented by $(\Sigma;R)$ is not left-cancellative, then there exist $a \in \Sigma$ and $y_1,y_2 \in \mathrm{Irr}(R)$ such that
(i) $y_1 \neq y_2$,
(ii) $ay_1 \overset{*}{\underset{R}{\leftrightarrow}} ay_2$,
(iii) no rule of R can be applied to both ay_1 and ay_2.

Proof. Assume that M is not left-cancellative, and let S consist of all pairs (x,y) of words from Σ^* such that $x \overset{*}{\underset{R}{\not\leftrightarrow}} y$ but $ax \overset{*}{\underset{R}{\leftrightarrow}} ay$ for some letter $a \in \Sigma$. By Lemma 5.2, the set S is non-empty. So let (y_1,y_2) be a pair from S such that $|y_1| + |y_2|$ is minimal, and let a be a corresponding letter from Σ.

Clearly, this means that y_1 and y_2 are irreducible, and hence, conditions (i) and (ii) are satisfied. If either ay_1 or ay_2 is also irreducible, then condition (iii) is satisfied as well, and we are done. Finally, assume that ay_1 and ay_2 can be reduced by applying the same rule (ℓ,r) of R. Then $\ell = ax$, $y_1 = xz_1$, and $y_2 = xz_2$, implying $ay_1 = axz_1 = \ell z_1 \underset{R}{\rightarrow} rz_1$ and $ay_2 = axz_2 = \ell z_2 \underset{R}{\rightarrow} rz_2$. Since $ay_1 \overset{*}{\underset{R}{\leftrightarrow}} ay_2$, we also have $rz_1 \overset{*}{\underset{R}{\leftrightarrow}} rz_2$, while as suffixes of y_1 and y_2, respectively, z_1 and z_2 are irreducible with $z_1 \neq z_2$, i.e., $z_1 \overset{*}{\underset{R}{\not\leftrightarrow}} z_2$. Hence, r can be factored as $r = sbt$, where $s,t \in \Sigma^*$ and $b \in \Sigma$, such that $|t|$ is minimal satisfying $tz_1 \overset{*}{\underset{R}{\not\leftrightarrow}} tz_2$ and $btz_1 \overset{*}{\underset{R}{\leftrightarrow}} btz_2$. This implies $(tz_1,tz_2) \in S$. On the other hand, $|tz_i| < |rz_i| < |xz_i| = |y_i|$ for

$i = 1,2$, and thus, $(tz_1, tz_2) \in S$ contradicts the choice of $(y_1, y_2) \in S$. Hence, no rule of R can be applied to both ay_1 and ay_2. □

Obviously, the condition given in Lemma 5.4 is not only necessary, but it is also sufficient for M being not left-cancellative.

Lemma 5.5. Let R be a length-reducing complete reduction system on Σ. Then the following two statements are equivalent:

(i) the monoid M presented by $(\Sigma; R)$ is not left-cancellative;

(ii) there exist a letter $a \in \Sigma$, and words $\ell, r \in \Sigma^*$ such that

 (a) $(a\ell, r) \in R$, and

 (b) $L(R_1, R) \cap L(R_2, R) \neq \emptyset$, where R_1 and R_2 are subsets of Σ^* defined as follows:

$$R_1 = a \cdot (\mathrm{Irr}(R) \cap \ell \cdot \Sigma^*), \text{ and}$$
$$R_2 = a \cdot (\mathrm{Irr}(R) \cap (\Sigma^* - \ell \cdot \Sigma^*)).$$

Proof. (i) ⇒ (ii): Assume that the monoid M is not left-cancellative. Then by Lemma 5.4 there are a letter $a \in \Sigma$ and irreducible words $y_1, y_2 \in \Sigma^*$ such that $y_1 \neq y_2$, $ay_1 \overset{*}{\underset{R}{\leftrightarrow}} ay_2$, and no rule of R can be applied to both ay_1 and ay_2. Without loss of generality we may assume that $|y_1| > |y_2|$. Since R is complete, we can conclude that ay_1 and ay_2 have a common descendant modulo R, which implies that ay_1 is reducible modulo R. Hence, there is a rule $(a\ell, r) \in R$ such that $y_1 = \ell u$ for some $u \in \Sigma^*$, and so we see that $ay_1 \in R_1 :=$ $a \cdot (\mathrm{IRR}(R) \cap \ell \cdot \Sigma^*)$. Since no rule of R can be applied to both ay_1 and ay_2, ℓ is not a prefix of y_2, and so $ay_2 \in R_2 := a \cdot (\mathrm{Irr}(R) \cap (\Sigma^* - \ell \cdot \Sigma^*))$. Thus, we see that $L(R_1, R) \cap L(R_2, R) \neq \emptyset$.

(ii) ⇒ (i): Let $a \in \Sigma$ and $\ell, r \in \Sigma^*$ such that $(a\ell, r) \in R$ and $L(R_1, R) \cap L(R_2, R) \neq \emptyset$, where $R_1 = a \cdot (\mathrm{Irr}(R) \cap \ell \cdot \Sigma^*)$ and $R_2 =$ $a \cdot (\mathrm{Irr}(R) \cap (\Sigma^* - \ell \cdot \Sigma^*))$. Then there are words $ay_1 \in R_1$ and $ay_2 \in R_2$ such that ay_1 and ay_2 have a common descendant, i.e., $ay_1 \overset{*}{\underset{R}{\leftrightarrow}} ay_2$. Since $ay_1 \in R_1$, we see that $y_1 \in \mathrm{Irr}(R)$ and $y_1 = \ell u$ for some $u \in \Sigma^*$. Since $ay_2 \in R_2$, we see that $y_2 \in \mathrm{Irr}(R)$, and that ℓ is not a prefix of y_2. Hence, $y_1 \neq y_2$, and so $y_1 \overset{*}{\underset{R}{\not\leftrightarrow}} y_2$. Hence, the monoid M is not left-cancellative. □

Narendran and O'Dunlaing [Na-O'Dun] have shown that it is undecidable in general whether or not a monoid presented by a finite length-reducing complete reduction system is left-cancellative. On the other hand, if R is a finite monadic complete reduction system on Σ, then for each rule $(a\ell, r) \in R$, the sets R_1 and R_2 are regular subsets of Σ^*, and finite state acceptors recognizing these sets can be constructed effectively (Theorem 3.1). But then finite state acceptors for $L(R_1,R)$ and $L(R_2,R)$ can be obtained (Theorem 3.2), implying that condition (ii) of Lemma 5.5 becomes decidable in this situation. Thus, we have shown the following.

Theorem 5.6 ([Na-O'Dun]).
The following problem is decidable:

INSTANCE: A finite presentation $(\Sigma;R)$, where R is a monadic complete reduction system on Σ.

QUESTION: Is the monoid M_R given through this presentation left-cancellative ?

As remarked earlier, a corresponding result also holds for right-cancellativity. Combining these results we obtain the following.

Corollary 5.7. The following problem is decidable:

INSTANCE: A finite presentation $(\Sigma;R)$, where R is a monadic complete reduction system on Σ.

QUESTION: Is the monoid M_R cancellative ?

Next we want to deal with the problem of the existence of non-trivial idempotents of a monoid given through a finite monadic complete reduction system.

Definition 5.8. Let M be a monoid presented by $(\Sigma;R)$. Then a word $u \in \Sigma^*$ describes an idempotent of M, if $u^2 \overset{*}{\underset{R}{\leftrightarrow}} u$. If, in addition, $u \overset{*}{\underset{R}{\not\leftrightarrow}} e$, then u describes a non-trivial idempotent of M.

As we saw in Section 2, it is undecidable in general whether or not a monoid M given through a finite presentation $(\Sigma;R)$ contains non-trivial idempotents. If, however, the reduction system R is monadic and complete, then those words that describe non-trivial idempotents can be characterized as follows.

Lemma 5.9. Let R be a monadic complete reduction system on Σ, and let $u \in \Sigma^*$ be a non-empty irreducible word. Then the following two statements are equivalent:

(i) u describes a non-trivial idempotent of the monoid M_R;

(ii) (a) $\exists x,y \in \Sigma^*$: $u = xy$ and $yx \overset{*}{\underset{R}{\rightarrow}} e$,

or

(b) $\exists x,y \in \Sigma^*$, $a \in \Sigma$: $u = xay$, $ayx \overset{*}{\underset{R}{\not\rightarrow}} e$, and $ayxa \overset{*}{\underset{R}{\rightarrow}} a$.

Proof. If u = xy with $yx \overset{*}{\underset{R}{\Rightarrow}} e$, then $u^2 = xyxy \overset{*}{\underset{R}{\Rightarrow}} xy = u$, i.e., u describes an idempotent. If u = xay with $ayx \overset{*}{\underset{R}{\not\Rightarrow}} e$ and $ayxa \overset{*}{\underset{R}{\Rightarrow}} a$, then $u^2 = xayxay \overset{*}{\underset{R}{\Rightarrow}} xay = u$, i.e., u again describes an idempotent. By hypothesis u is a non-empty irreducible word implying that $u \overset{*}{\underset{R}{\not\leftrightarrow}} e$. Thus, u describes a non-trivial idempotent of the monoid M_R.

On the other hand, if u describes an idempotent, then $u^2 \overset{*}{\underset{R}{\leftrightarrow}} u$ implying that $u^2 \overset{*}{\underset{R}{\Rightarrow}} u$ according to the choice of u. Since R is monadic, there are words $x,x',y,y' \in \Sigma^*$ and some $a \in \Sigma \cup \{e\}$ such that $u = xx' = y'y = xay$ with $x'y' \overset{*}{\underset{R}{\Rightarrow}} a$. Hence, $x' = ay$ and $y' = xa$.

If a = e, then we have u = xy with $yx = x'y' \overset{*}{\underset{R}{\Rightarrow}} e$, i.e., (ii) (a) is satisfied. If $a \in \Sigma$, then u = xay with $ayxa = x'y' \overset{*}{\underset{R}{\Rightarrow}} a$. If $ayx \overset{*}{\underset{R}{\not\Rightarrow}} e$, then (ii) (b) is satisfied, otherwise u = x(ay) with $(ay)x \overset{*}{\underset{R}{\Rightarrow}} e$, i.e., (ii) (a) is satisfied. □

This result will be useful for deriving a necessary and sufficient condition for a monoid M presented by a monadic complete reduction system to contain non-trivial idempotents. However, before deriving this condition, we still need the following notion.

Definition 5.10. Let R be a monadic complete reduction system on Σ. Then for $a \in \Sigma$, $INT_R(a) := \{w \in Irr(R) \mid awa \overset{*}{\underset{R}{\Rightarrow}} a\}$ is the set of internal factors of the letter a.

Lemma 5.11. Let R be a monadic complete reduction system on Σ, and let $a \in \Sigma$. If $|INT_R(a)| \geq 2$, then the monoid M_R does have a non-trivial idempotent.

Proof. Let $a \in \Sigma$ with $|INT_R(a)| \geq 2$, and let $w_1, w_2 \in INT_R(a)$ with $w_1 \neq w_2$. Then for i = 1,2, we have $aw_ia \overset{*}{\underset{R}{\Rightarrow}} a$. If $aw_i \overset{*}{\underset{R}{\not\Rightarrow}} e$, then aw_i describes a non-trivial idempotent of M_R, if $w_ia \overset{*}{\underset{R}{\not\Rightarrow}} e$, then w_ia describes a non-trivial idempotent of M_R. So assume that $aw_i \overset{*}{\underset{R}{\Rightarrow}} e$ and $w_ia \overset{*}{\underset{R}{\Rightarrow}} e$ for i = 1,2. Then $w_1 \overset{*}{\underset{R}{\leftrightarrow}} w_1aw_2 \overset{*}{\underset{R}{\leftrightarrow}} w_2$ implying $w_1 = w_2$, since w_1 and w_2 are irreducible, and R is complete. But this contradicts the choice of w_1 and w_2. □

Now we can characterize the existence of non-trivial idempotents in monoids presented by monadic complete reduction systems as follows.

Theorem 5.12. Let R be a monadic complete reduction system on Σ. Then the monoid M_R contains a non-trivial idempotent if and only if one of the following three conditions is satisfied:

(i) $\exists u \in Irr(R)-\{e\}\ \exists x,y \in \Sigma^*$: $u = xy$ and $yx \overset{*}{\underset{R}{\Rightarrow}} e$, or

(ii) $\exists a \in \Sigma$: $|INT_R(a)| \geq 2$, or

(iii) $\exists a \in \Sigma\ \exists w \in Irr(R)$: $INT_R(a) = \{w\}$ and $aw \overset{*}{\underset{R}{\not\Rightarrow}} e$.

Proof. If there exists a word u $\in$ Irr(R)-{e} satisfying (i), then this word u describes a non-trivial idempotent of M_R by Lemma 5.9. By Lemma 5.11 (ii) implies that M_R contains a non-trivial idempotent. Finally, if $\mathrm{INT}_R(a)$ = {w} with $aw \not\xrightarrow[R]{*} e$ for some a $\in \Sigma$, then u = aw describes a non-trivial idempotent of M_R.

So assume conversely, that M_R contains a non-trivial idempotent, i.e., there exists a word u $\in$ Irr(R)-{e} describing a non-trivial idempotent of M_R. Then according to Lemma 5.9 u satisfies condition (i), or there are words x,y $\in \Sigma^*$ and a letter a $\in \Sigma$ such that u = xay, $ayx \not\xrightarrow[R]{*} e$, and $ayxa \xrightarrow[R]{*} a$. Take w $\in \Sigma^*$ to be the irreducible descendant of yx modulo R. Then $aw \not\xrightarrow[R]{*} e$, and $awa \xrightarrow[R]{*} a$ implying that w $\in \mathrm{INT}_R(a)$. Thus, either condition (ii) is satisfied with a, or condition (iii) is satisfied with a and w. □

Thus, in order to determine whether a monoid M given by a monadic complete reduction system R on Σ contains a non-trivial idempotent, it is sufficient to check conditions (i) to (iii) of Theorem 5.12.

Define an operation CYCLE on the set $P(\Sigma^*)$ of languages on Σ by taking CYCLE(L) = {yx | xy $\in$ L}, i.e., CYCLE(L) is the language that contains all the cyclic permutations of words of L. By using this operation we can express condition (i) of Theorem 5.12 as follows:

(i') e $\in$ L(CYCLE(Irr(R)-{e}),R).

Given a finite monadic complete reduction system R on Σ, we can effectively construct a finite state acceptor A_1 for recognizing the set Irr(R)-{e} (Theorem 3.1). By using well-known techniques from automata theorey (c.f., e.g., [Ho-Ul]), we then obtain a finite state acceptor A_2 for the set CYCLE(Irr(R)-{e}), from which we finally derive a finite state acceptor A_3 for the set L(CYCLE(Irr(R)-{e}),R) (Theorem 3.2). Thus, we have the following result.

Lemma 5.13. The following problem is decidable:

INSTANCE: A finite monadic complete reduction system R on Σ.

QUESTION: Does there exist a word u $\in$ Irr(R)-{e} that can be factored as u = xy with $yx \xrightarrow[R]{*} e$?

Lemma 5.13 shows that condition (i) of Theorem 5.12 is decidable.

Lemma 5.14. Let R be a finite monadic complete reduction system on Σ. Then for each a $\in \Sigma$, one can effectively construct a deterministic pushdown automaton M_a that recognizes the set $\mathrm{INT}_R(a)$. In particular, $\mathrm{INT}_R(a)$ is a deterministic context-free language for each a $\in \Sigma$.

Proof. Given a finite monadic complete reduction system R on Σ and a letter $a \in \Sigma$, one can effectively construct a deterministic pushdown automaton (dpda) M_1 recognizing the set $[a]_R$ and a finite state acceptor M_2 recognizing the set $a \cdot Irr(R) \cdot a$ (Theorems 3.1 and 3.3). From M_1 and M_2 we get a dpda M_3 that recognizes the set $[a]_R \cap a \cdot Irr(R) \cdot a = \{awa \mid w \in Irr(R) \text{ with } awa \overset{*}{\underset{R}{\leftrightarrow}} a\}$. Since R is complete, we have $awa \overset{*}{\underset{R}{\leftrightarrow}} a$ if and only if $awa \overset{*}{\underset{R}{\rightarrow}} a$. Thus, $[a]_R \cap a \cdot Irr(R) \cdot a = \{awa \mid w \in Irr(R) \text{ with } awa \overset{*}{\underset{R}{\rightarrow}} a\} = a \cdot INT_R(a) \cdot a$. By using well-known techniques for dealing with dpda's (cf., e.g., [Ha 78]) a dpda M_4 recognizing the set $INT_R(a)$ can effectively be constructed from M_3. □

Given a deterministic pushdown automaton M we can determine the cardinality of the set L recognized by M, and if L is finite, we can compute a list containing all the elements of L. Hence, conditions (ii) and (iii) of Theorem 5.12 are also decidable for finite monadic complete reduction systems. Thus, we have the following result.

Theorem 5.15 ([Ot 85a]). The following problem is decidable:

INSTANCE: A finite presentation $(\Sigma; R)$, where R is a monadic complete reduction system on Σ.

QUESTION: Does the monoid M_R presented by $(\Sigma; R)$ contain any non-trivial idempotents ?

In the proofs of Lemmas 5.13 and 5.14 an algorithm is developed for solving the above decision problem. Since given a finite monadic complete reduction system R on Σ, this algorithm at first derives a finite state acceptor for the set $Irr(R)$, it needs in general exponential time and space. It is not yet known whether there exists a more efficient algorithm for solving this problem. If finite length-reducing complete reduction systems are considered that are not monadic, then the technique used to develop the above algorithm is not applicable any more. So the question remains open of whether or not the above problem can be solved when being considered for presentations involving finite length-reducing complete reduction systems that are not monadic.

Now we turn to the problem of deciding of whether or not a monoid given through a finite monadic complete reduction system contains any non-trivial elements of finite order.

Definition 5.16. Let M be a monoid presented by $(\Sigma; R)$. Then a word $w \in \Sigma^*$ describes an element of finite order of M, if there exist integers $n > 0$ and $k > 1$ such that $w^{n+k} \overset{*}{\underset{R}{\leftrightarrow}} w^n$. If, in addition, $w \overset{*}{\underset{R}{\not\leftrightarrow}} e$, then w is said to describe a <u>non-trivial element of finite order</u> of M.

As we saw in Section 2, it is undecidable in general whether or not a finitely presented monoid contains non-trivial elements of finite order. On the other hand, Lallement has given a syntactic characterization for those one-rule reduction systems $R = \{(\ell,r)\}$ on Σ, for which the monoid M_R does contain such elements [La 74]. Since it is decidable whether or not a one-rule reduction system $R = \{(\ell,r)\}$ meets this syntactic characterization, the above decision problem is decidable for monoids given through one-rule reduction systems. Here we want to extend this result to the class of all monoids that are presented by finite monadic complete reduction systems.

So let R be a finite monadic complete reduction system on Σ. According to the results presented before it is decidable whether or not the monoid M_R presented by $(\Sigma;R)$ contains a non-trivial idempotent. If M_R contains a non-trivial idempotent, then this idempotent is a non-trivial element of finite order of M_R. However, M_R may have non-trivial elements of finite order, although it has no non-trivial idempotents. For example, this situation occurs when the monoid M_R is a finite group. Therefore, we must derive additional information on words describing non-trivial elements of finite order of the monoid M_R in case M_R does not contain any non-trivial idempotents. The following lemma provides us with the necessary information.

<u>Lemma 5.17</u>. Let R be a finite length-reducing complete reduction system on Σ, and let $\mu = \max\{|\ell| \mid \ell \in \text{left}(R)\}$. If the monoid M_R presented by $(\Sigma;R)$ does not contain any non-trivial idempotents, then the following two statements are equivalent:

(i) The monoid M_R contains a non-trivial element of finite order.

(ii) There is a word $w \in \Sigma^*$ of length $|w| < \mu$ such that w describes a non-trivial element of finite order of M_R.

<u>Proof</u>. Obviously it suffices to prove that (i) implies (ii). So let $w \in \Sigma^*$ be a shortest word describing a non-trivial element of finite order of M_R. If $|w| < \mu$, then we are done. Hence, assume that $|w| > \mu$. Since w describes a non-trivial element of finite order of M_R, there exist integers $n > 0$ and $k > 1$ such that $w^{n+k} \overset{*}{\underset{R}{\leftrightarrow}} w^n$, and $w \overset{*}{\underset{R}{\not\leftrightarrow}} e$. This implies in particular that $n+k > 2$. Since R is complete, the words w^{n+k} and w^n have a common descendant modulo R, which means that w^{n+k} is reducible modulo R. But $|w| > \mu$, and so w^2 is reducible modulo

R. Thus, $w = w_1w_2 = w_3w_4$ with $(w_2w_3,r) \in R$ for some words w_1,w_2,w_3,w_4, $r \in \Sigma^*$ with $|r| < |w_2w_3|$.

Now $|w_2w_3| < \mu < |w| = |w_1w_2|$ implying that $|w_3| < |w_1|$. Hence, there is a word $x \in \Sigma^*$ such that $w_1 = w_3x$, i.e., $w = w_3xw_2$. This gives: $(xr)^{n+k+1} \overset{*}{\underset{R}{\longleftrightarrow}} (xw_2w_3)^{n+k+1} = xw_2(w_3xw_2)^{n+k}w_3 = xw_2w^{n+k}w_3 \overset{*}{\underset{R}{\longleftrightarrow}} xw_2w^nw_3 = (xw_2w_3)^{n+1} \overset{*}{\underset{R}{\longleftrightarrow}} (xr)^{n+1}$. If $xr \not\overset{*}{\underset{R}{\longleftrightarrow}} e$, this shows that xr does describe a non-trivial element of finite order of M_R, although $|xr| < |xw_2w_3| = |w|$, thus contradicting the choice of w. On the other hand, if $xr \overset{*}{\underset{R}{\longleftrightarrow}} e$, then we have the following: $w^2 = (w_3xw_2)^2 = w_3(xw_2w_3)xw_2 \underset{R}{\longleftrightarrow} w_3(xr)xw_2 \overset{*}{\underset{R}{\longleftrightarrow}} w_3xw_2 = w$, i.e., w describes a non-trivial idempotent of M_R, thus contradicting the assumptions about M_R. □

So in order to decide whether or not a monoid M given through a finite monadic complete reduction system R on Σ contains any non-trivial elements of finite order, we can proceed as follows. At first we determine whether or not M contains any non-trivial idempotents. If it does, then the answer to our question is "yes". If it does not, then we enumerate all the words from Σ^* of length up to $\mu = \max\{|\ell| \mid \ell \in \text{left}(R)\}$ and test whether any of these words describe non-trivial elements of finite order of M. The following lemma shows that this test can be performed effectively.

<u>Lemma 5.18</u>. The following problem is decidable:

INSTANCE: A finite monadic complete reduction system R on Σ, and a word $w \in \Sigma^*$.

QUESTION: Does w describe a non-trivial element of finite order of the monoid M_R ?

<u>Proof</u>. Given R and w, we can easily decide whether or not $w \overset{*}{\underset{R}{\longleftrightarrow}} e$ holds. If $w \overset{*}{\underset{R}{\longleftrightarrow}} e$, then obviously w does not describe a non-trivial element of finite order of M_R. So assume that $w \not\overset{*}{\underset{R}{\longleftrightarrow}} e$. From w we can construct a nondeterministic finite state acceptor A_w for the language $\{w\}^+$. From A_w and R we get a nondeterministic finite state acceptor A_w^Δ for the language $L(\{w\}^+,R)$ and a finite state acceptor A_R for the set $\text{Irr}(R)$. By combining A_w^Δ and A_R we finally get a nondeterministic finite state acceptor $A(R,w)$ that recognizes the language $L(\{w\}^+,R) \cap \text{Irr}(R) = \{u \in \text{Irr}(R) \mid$ There exists an integer $k \geq 1$ such that $w^k \overset{*}{\underset{R}{\rightarrow}} u\}$. It is obvious that w describes an element of finite order of M_R if and only if this language is finite. But the cardinality of this language can be determined effectively from $A(R,w)$. □

Now Theorem 5.15, Lemma 5.17, and Lemma 5.18 together yield the following result.

Theorem 5.19 ([Ot 85a]). The following problem is decidable:

INSTANCE: A finite presentation $(\Sigma;R)$, where R is a monadic complete reduction system on Σ.

QUESTION: Does the monoid M_R presented by $(\Sigma;R)$ contain any non-trivial elements of finite order ?

Since a monoid that does not contain any non-trivial elements of finite order is called torsion-free, the above result can also be stated as follows.

Corollary 5.20. The following problem is decidable:

INSTANCE: A finite presentation $(\Sigma;R)$, where R is a monadic complete reduction system on Σ.

QUESTION: Is the monoid M_R presented by $(\Sigma;R)$ torsion-free ?

The algorithm developed above for deciding torsion-freeness of monoids given through finite monadic complete reduction systems needs exponential time and space in general. So far it is not known whether there exists a more efficient algorithm for solving this problem. However, when we restrict our attention to cancellative monoids that are presented by finite monadic complete reduction systems, then those monoids that are not torsion-free can be recognized non-deterministically in polynomial time, since a cancellative monoid does not contain any non-trivial idempotents (for details cf. [Ot 85b]). On the other hand, it is an open problem of whether or not the property of torsion-freeness is decidable for monoids presented by finite length-reducing complete reduction systems that are not monadic, although Lemma 5.17 still holds in this situation.

Theorem 4.12 and Corollary 5.20 can now be combined to show that it is decidable whether or not a monoid M given through a finite monadic complete reduction system R on Σ is a free group. Recall that a monoid M is a free group if and only if it has a presentation of the form $(\Sigma \cup \bar{\Sigma}; \{(a\bar{a},e),(\bar{a}a,e) \mid a \in \Sigma\})$, where $\bar{\Sigma}$ is an alphabet in 1-to-1 correspondence with Σ, $\Sigma \cap \bar{\Sigma} = \emptyset$, and $\bar{\ }: \Sigma \to \bar{\Sigma}$ is a bijection realizing this correspondence.

Let R be a finite monadic complete reduction system on Σ. Then each congruence class $[u]_R$ is a deterministic context-free language. In particular, $[e]_R$ is a context-free language. So if the monoid M

presented by $(\Sigma;R)$ happens to be a group, then it is a context-free group. Hence, the following result of Muller and Schupp is applicable.

Theorem 5.21 ([Mu-Schu]).
A finitely generated torsion-free group G is free if and only if it is context-free.

Since each free group is torsion-free, this implies the following.

Corollary 5.22. A finitely generated context-free group is free if and only if it is torsion-free.

Thus, the monoid M presented by $(\Sigma;R)$ is a free group if and only if it is a group and it is torsion-free. Thus, by combining the corresponding decidability results we obtain the following.

Theorem 5.23. The following problem is decidable:

INSTANCE: A finite presentation $(\Sigma;R)$, where R is a monadic complete reduction system on Σ.

QUESTION: Is the monoid M_R given through this presentation a free group ?

Finally we want to look at the existence of elements of infinite order.

Definition 5.24. Let M be a monoid presented by $(\Sigma;R)$. Then a word $w \in \Sigma^*$ describes an element of infinite order of M if $w^i \not\overset{*}{\underset{R}{\leftrightarrow}} w^j$ for all $i,j \in \mathbb{N}$, $i \neq j$.

If R is a finite complete reduction system on Σ, then we can decide effectively whether or not the monoid M presented by $(\Sigma;R)$ is finite. Obviously, if M is finite, then all its elements have finite order. But does also the reverse implication hold ?

In general this problem is known as the Burnside problem. It has been shown that there exist infinite monoids all the elements of which have finite order, i.e., the reverse implication mentioned above does not hold in general. In fact, Adjan [Ad 79] has proven that for each integer $m > 1$ and each odd integer $n \geq 665$, the group $B(m,n)$ presented by $(\Sigma \cup \bar{\Sigma};\ \{(a\bar{a},e),(\bar{a}a,e) \mid a \in \Sigma\} \cup \{(x^n,e) \mid x \in (\Sigma \cup \bar{\Sigma})^*\})$, where $\Sigma = \{a_1,a_2,\ldots,a_m\}$ and $\bar{\Sigma} = \{\bar{a}_1,\bar{a}_2,\ldots,\bar{a}_m\}$, is infinite. All these groups are finitely generated only, they are not finitely presented [Ad 79]. This leaves the problem of whether or not there exists a finitely presented infinite group all of which elements have finite order. Here, we will show that such a group, if it should exist, cannot be presented by a finite complete reduction system.

Lemma 5.25. Let R be a finite complete reduction system on Σ such that the monoid M_R presented by $(\Sigma;R)$ is infinite. Then this monoid contains an element of infinite order.

Proof. Since R is a complete reduction system on Σ, the set Irr(R) of irreducible words modulo R is a set of representatives for the monoid M_R. Thus, with M_R also Irr(R) is infinite. Since R is finite, the set Irr(R) is a regular subset of Σ^*. Hence, by the pumping lemma of regular sets, there exists an integer $n > 1$ such that each word $u \in \mathrm{Irr}(R)$ of length $|u| > n$ can be factored as $u = xyz$, where $y \neq e$ and $\{xy^kz \mid k > 0\} \subseteq \mathrm{Irr}(R)$. Since Irr($R$) is infinite, there exists a word $u \in \mathrm{Irr}(R)$ of length $|u| > n$. Then $u = xyz$ satisfying $y \neq e$ and $\{xy^kz \mid k > 0\} \subseteq \mathrm{Irr}(R)$. But the set Irr($R$) is subword-closed implying that $\{y^k \mid k > 0\} \subseteq \mathrm{Irr}(R)$. Since R is complete we can conclude that $y^i \not\overset{*}{\leftrightarrow}_R y^j$ for all $i,j \in \mathbb{N}$ with $i \neq j$. Thus, y describes an element of infinite order of M_R. □

Hence, when a monoid M is presented by a finite complete reduction system, then this monoid contains an element of infinite order if and only if it is an infinite monoid. Thus, Theorem 4.1 immediately gives the following corollary.

Corollary 5.26. The following problem is decidable:

INSTANCE: A finite presentation $(\Sigma;R)$, where R is a complete reduction system on Σ.

QUESTION: Does the monoid M_R presented by $(\Sigma;R)$ contain an element of infinite order ?

We have seen that all of the ten decision problems listed at the beginning of Section 2 become decidable, when they are restricted to monoids presented by finite complete reduction systems or certain specializations thereof. Clearly, this observation raises the question of which monoids and groups can at all be presented in this way. More formally, is there an algebraic characterization for these monoids and groups ? This fundamental question is still unanswered, so far only a few partial results could be obtained. Cochet [Co 76] proved that a group G can be described by a presentation $(\Sigma;R)$, where R is a finite special complete reduction system on Σ, if and only if G is the free product of finitely many cyclic groups. Avenhaus, Madlener and Otto [Av-Ma-Ot] proved that a group G can be described by a presentation $(\Sigma;R)$, where R is a finite two-monadic complete reduction system on Σ, if and only if G is the free product of a finitely generated free group and a finite number of finite groups. Here, a monadic reduction system R is called <u>two-monadic</u>, if each rule of R has a left-hand side of length two. It has been conjectured (cf., e.g., [Gil 84]) that the same algebraic characterization holds for the class of groups presented by finite monadic complete reduction systems, but so far this has not been shown.

We close this chapter with a short list containing some of the most interesting open problems in connection with the material presented.

(i) Is it possible to extend Theorems 5.15, 5.19, and 5.23 to the class of all monoids presented by finite length-reducing complete reduction systems ? Is it possible to extend these results even further ?

(ii) Does Gilman's conjecture hold ?

(iii) Which groups can be presented by finite length-reducing complete reduction systems ?

(iv) Which groups can be presented by finite complete reduction systems ?

(v) Which monoids can be presented by finite (special, monadic, length-reducing) complete reduction systems ? Note that it has been shown only recently by Squier [Sq-1] that there exists a finitely presented monoid with a decidable word problem such that this monoid cannot be presented by a finite complete reduction system.

References:

[Ad 55] **S.I. Adjan:**
The Algorithmic Solvability of Problems Concerning Certain Properties of Groups
Dokl. Akad. Nauk SSSR 103 (1955), 533-535

[Ad 79] **S.I. Adjan:**
The Burnside Problem and Identities in Groups
Springer Verlag, Berlin, Heidelberg, New York 1979

[Av-Ma-Ot] **J. Avenhaus, K. Madlener, F. Otto:**
Groups Presented by Finite Two-monadic Church-Rosser Thue Systems
Transactions Americ. Math. Soc., to appear

[Ba-Bu-La 84] **M. Ballantyne, G. Butler, D. Lankford:**
Three Problems in Applied Communtative Logic
Notes, 1984

[Ba-La 81] **M. Ballantyne, D. Lankford:**
New Decision Algorithms for Finitely Presented Semigroups
J. Comp. and Math. with Appl. 7 (1981), 159-165

[Ba 75] **H. Bass:**
The Degree of Polynomial Growth of Finitely Generated Nilpotent Groups
Proc. London Math. Soc. (3) 25 (1975), 603-614

[Ba-Ot] **G. Bauer, F. Otto:**
Finite Complete Rewriting Systems and the Complexity of the Word Problem
Acta Informatica 21 (1984), 521-540

[Bl 77] **W.W. Bledsoe:**
The SUP-INF Method in Presburger Arithmetic,
Memo ATP-18, Austin, Texas 1979

[B-D-J 78] **D. Brand, J. Darringer, J. Joyner:**
Completeness of Conditional Reductions
Proc. of the Fourth Workshop on Automated Deduction
Austin, Texas (1979), 36-43

[Bo 82] **R.V. Book:**
When is a Monoid Group? The Church-Rosser Case is Tractable
Theoret. Comput. Sci. 18 (1982), 325-331

[Bo 83] **R.V. Book:**
Decidable Sentences of Church-Rosser Congruences
Theoret. Comput. Sci. 24 (1983), 301-312

[Bo 84] **R.V. Book:**
String Rewriting Systems with Applications to Algebraic Protocols
Lectures Notes of a Course Given at the University of Stuttgart, November 1984

[Bo-Ja-Wr 82] **R.V. Book, M. Jantzen, C. Wrathall:**
Monadic Thue Systems
Theoret. Comput. Sci. 19 (1982), 231-251)

[Bo-Ot 85] **R.V. Book, F. Otto:**
Cancellation Rules and Extented Word Problems
Inf. Proc. Letters 20 (1985), 5-11

[Boo-Ro 66] **W.W. Boone, H. Rogers jr.:**
On a Problem of J.H.C. Whitebread and a Problem of Alonzo Church
Math. Scand. 19 (1966), 185-192

[Br 64] **J.A. Brozozowski:**
Derivatives of Regular Expressions
Journal of the ACM 11 (1964), 481-494

[Bu 65] **B. Buchberger:**
Ein Algorithmus zum Auffinden der Basis-elemente des Restklassenrings nach einem nulldimensionalen Polynomideal
Dissertation, Innsbruck 1965

[Bu 83] **B. Buchberger:**
A critical Pair Completion Algorithm for Finitely Generated Ideals in Rings
In: Logic and Machines: Decision Problems and Complexity, ed. E. Börger et al., LNCS 171 (1984), 137-161

[Bü 79] **H. Bücken:**
Anwendungen von Reduktionsystemen auf das Wortproblem in der Gruppentheorie
Dissertation, Aachen 1979
(also: Proc. of the Fourth Workshop of Automated Deduction, Austin, Texas 1979)

[C-M 80] **H.S.M. Coxeter, W.O. Moser:**
Generators and Relations for Discrete Groups
Springer Verlag, Berlin, Heidelberg, New York, 1980

[Co 76] **Y. Cochet:**
Church-Rosser Congruences on Free Semigroups
Coll. Math. Soc. Janos Bolyai: Algebraic
Theory of Semigroups 20 (1976), 51-60

[Da 58] **M. Davis:**
Computability and Unsolvability
McGraw Hill, 1958

[De 11] **M. Dehn:**
Über unendliche diskontinuierliche Gruppen
Math. Ann. 71 (1911), 116-144

[Der 82] **N. Dershowitz:**
Orderung for Term Rewriting Systems
J. Theor. Comp. Sci. 17 (1982), 279-301

[Der 85] **N. Dershowitz:**
Termination of Rewriting
Report No. UIUCDS-R-85-1220, Dept. of Comp. Sc., Univ. of Ill. at Urbana-Champaign, Aug. 1985

[Der-Ma 79] **N. Dershowitz, Z. Manna:**
Proving Termination with Multiset Orderings
Comm. ACM 22 (1979), 465-476

[vdD-Wi 84] **L. van den Dries, A.J. Wilkie:**
On Gromov's Theorem on Groups of Polynomial Growth and Elementary Logic
Journal of Algebra 89 (1984), 349-374

[Ev 51] **T. Evans:**
On Muliplicative Systems Defined by Generators and Relations. I. Normal Forms Theorems
Proc. Cambridge Philos.Soc. 47 (1951), 637-649

[Ger 70] **J.A. Gerhardt:**
The Lattice of Equational Classes of Idempotent Semigroups
Journ. of Algebra 15 (1970)

[Gre 60] **M. Greendlinger:**
Dehn's Algorithm for the Word Problem
Comm. Pure Appl. Math. 13 (1960), 641-677

[Gre-Re 52] **J.H. Green, D. Rees:**
On Semigroups in which $x^r = x$
Proc. Cambridge Phil. Soc. 48 (1952), 35-40

[Gi 79] **R.H. Gilman:**
Presentations of Groups and Monoids
Journal of Algebra 57 (1979), 544-554

[Gil 84] **R.H. Gilman**
Computations with Rational Subsets of Confluent Groups
Proc. of EUROSAM 84, LNCS 174 (1984), 207-212

[Ha 78] **M.A. Harrison:**
Introduction to Formal Language Theory
Addison-Wesley, 1978

[H-N-N 49] **G. Higman, B.H. Neumann, H. Neumann:**
Embedding Theorems for Groups
J. London Math. Soc. 24 (1949), 247-254

[Ho 83] **P. Horster:**
Reduktionsysteme, formale Sprachen und Automatentheorie
Dissertation, Aachen, 1983

[Ho-Ul 79] **J.E.Hopcroft, J.D. Ullman:**
Introduction to Automata Theory, Languages and Computation
Addison-Wesley, 1979

[Hs 82] **J. Hsiang:**
Topics in Automated Theorem Proving and Program Generation
Report No. UIUCDCS-R-82-1113, Univ. of Ill. at Urbana-Champaign, Dec. 1982

[Hu 80] **G. Huet:**
Confluent Reductions; Abstract Properties and Applications to Term Rewriting Systems
Journal of the ACM 27, (1980), 797-821

[Hu-Op 80] **G. Huet, D. Oppen:**
Equations and Rewrite Rules - A Survey
In: Formal Language Theory, ed. R.V. Book, Acad. Press, New York (1980), 341-405

[Jou-Le-Re 82] **J.P. Jouannaud, P. Lescannes, F. Reinig:**
Recursive Decomposition Ordering
Proc. Second IFIP Workshop on Formal Description of Programming Concepts (1982), 331-348

[Ka-Kr-McN-Na 85] **D. Kapur, M. Krishnamoorthy, R. McNaughton, P. Narendran:**
An $o(|T|^3)$ Algorithm for Testing the Church-Rosser Property of Thue Systems
Theoret. Comput.Sci. 35 (1985), 109-114

[Ka-Na 85] **D. Kapur, P. Narendran:**
An Equational Approach to Theorem Proving in First-order Predicate Calculus
Report No. 84 CRD 322, General Electric, Sept. 1985

[Ka-Na-Si 85] **D. Kapur, P. Narendran, G. Sivakumar:**
A Path Ordering for Proving Termination of Term Rewriting Systems
Proc. Tenth Coll. on Trees in Algebra and Progamming, 1985

[Ka-Me 79] **M.I. Kargapolov, Y.I. Merzliakov:**
Fundamentals of the Theory of Groups
Springer Verlag, Berlin, Heidelberg, New York, 1979

[Ke 83] **S. Kemmerich:**
Unendliche Reduktionssysteme
Dissertation, Aachen, 1983

[Kn-Be 70] **D.E. Knuth, P.B.Bendix:**
Simple Word Problems in Universal Algebras
In: Computational Problems in Abstract Algebras, ed. H. Leech, Pergamon Press (1970), 263-297

[Kü 86] **W. Küchlin:**
Equational Completion by Proof Simplification
ETH Report 86-02, Zürich 1986

[La 74] **G. Lallement:**
On Monoids Presented by a Single Relation
Journal of Algebra 32 (1974), 370-388

[La 75] **D. Lankford:**
Canonical Inference
Memo ATP-32, Univ. of Texas, Austin, Dec. 1975

[La 79] **D. Lankford:**
On Proving Term Rewriting Systems are Noetherian
Memo MTP-3, Math. Dept. Louisiana, Tech. Univ., 1979

[La 81] **D. Lankford:**
Research in Applied Equational Logic
Reprint, Math. Dept. Louisiana, Tech. Iniv. (MTP-15), 1981

[Ly-Sch 77] **R.C. Lyndon, P.E. Schupp:**
Combinatorial Group Theory
Springer Verlag, Berlin, Heidelberg, New York, 1977

[Mag 32] **W. Magnus:**
Das Identitätsproblem bei Gruppen mit einer definierenden Relation
Math. Annalen 106 (1932), 295-307

[Ma- Ka-So 76] **W. Magnus, A. Karrass, D. Solitar:**
Combinatorial Group Theory
2nd. rev. ed., Dover, New York, 1976

[Ma 51] **A. Markov:**
Impossibility of Algorithms for Recognizing Some Properties of Associative Systems
Dokl. Akad. Nauk SSSR 77 (1951), 953-956

[Mo 52] **A. Mostowski:**
Review of [Ma 51]
J. Symbolic Logic 17 (1952), 151-152

[Mu-Schu 83] **D.E. Muller, P.E. Schupp:**
Groups, the Theory of Ends, and Context-free Languages
J. Comp. System Sci. 26 (1983), 295-310

[Na-O'Dun] **P. Narendran, C. O'Dunlaing:**
Cancellativity in Finitely Presented Semigroups
Submitted for Publication

[New 42] **M. Newman:**
On Theories with a Combinatorial Definition of Equivalence
Ann. of Math. 43 (1942), 223-243

[Ni 70] **M. Nivat:**
On some Families of Languages Related to the Dyck-Languages
Sec. ACM Syposium Theory Computing (1970), 221-225

[Nov 58] **P.S. Novikov:**
On the Algorithmic Unsolvability of the Word Problem in Group Theory
Amer. Math. Soc. Translation Series 2, Vol. 9 (1958), 1-122

[Ot] **F. Otto:**
Church-Rosser Thue Systems that Present Free Monoids
SIAM J. on Comp., to appear

[Ot 85a] **F. Otto:**
Elements of Finite Order for Finite Monadic Church-Rosser Thue Systems
Transactions of the American Math. Soc. 291 (1985), 629-637

[Ot 85b] **F. Otto:**
Decision Problems and their Complexity for Monadic Church-Rosser Thue Systems
Habilitationsschrift, Univ. Kaiserslautern, 1985

[Ot 86] **F. Otto:**
On Deciding whether a Monoid is a Free Monoid or is a Group
Acta Inform. 23 (1986), 99-110

[O'Dun 83] **C. O'Dunlaing:**
Infinite Regular Thue-systems
Theor. Comp. Science 25 (1983), 171-192

[Pe-Sti 77] **G. Peterson, M. Stickel:**
Complete sets of Reductions for Équational Theories with Complete Unification Algorithms
Tech. Report, Univ. of Arizona 1977

[Rab 58] **M. Rabin:**
Recursive Unsolvability of Group Theoretic Problems
Ann. of Math.67 (1958), 172-194

[Ri 78] **M.M. Richter:**
Logikkalküle, Teubner Verlag 1978

[Ri 82] **M.M. Richter:**
Complete and Incomplete Syystem of Reductions
In: Informatikfachberichte 57 (1982), ed. J. Nehmer, Springer Verlag

[Ro 65] **J.A. Robinson:**
A Machine-Oriented Logic Based on the Resolution Principle
JACM 12 (1965), 23-41

[Rot 73] **J. Rotman:**
The Theory of Groups
Boston 1973

[Sh 77] **R. Shostak:**
On the SUP-INF Method for Proving Presburger Formulas
J. ACM 24 (1977), 529-543

[Sie-Sz 83] **J. Siekmann, P. Szabo:**
A Noetherian and Confluent Rewrite System for Idempotent Semigroups
Semigroup Forum 25 (1983), 83-110

[Sq] **C. Squier:**
Word Problems and a Homological Finiteness Condition for Monoids
submitted for publication

[Sti 81] **M.E. Stickel:**
A Unification Algorithm for Associative-Commutative Functions.
J. ACM 28 (1981), 423-434

[Ti 08] **H. Tietze:**
Über die topologischen Invarianten mehrdimensionaler Mannigfaltigkeiten
Monatshefte für Mathematik und Physik 19 (1908), 1-118

[Wo 68] **J.A. Wolf:**
Growth of Finitely Generated Soluble Groups and Curvature of Riemannian Manifolds
J. Differential Geometry 2 (1968), 421-446

Subject Index

List of Symbols and Abbreviations

Lecture Notes in Computer Science

Vol. 192: Automata on Infinite Words. Proceedings, 1984. Edited by M. Nivat and D. Perrin. V, 216 pages.1985.

Vol. 193: Logics of Programs. Proceedings, 1985. Edited by R. Parikh. VI, 424 pages. 1985.

Vol. 194: Automata, Languages and Programming. Proceedings, 1985. Edited by W. Brauer. IX, 520 pages. 1985.

Vol. 195: H. J. Stüttgen, A Hierarchical Associative Processing System. XII, 273 pages. 1985.

Vol. 196: Advances in Cryptology. Proceedings of CRYPTO '84. Edited by G. R. Blakley and D. Chaum. IX, 491 pages. 1985.

Vol. 197: Seminar on Concurrency. Proceedings, 1984. Edited by S. D. Brookes, A. W. Roscoe and G. Winskel. X, 523 pages. 1985.

Vol. 198: A. Businger, PORTAL Language Description. VIII, 186 pages. 1985.

Vol. 199: Fundamentals of Computation Theory. Proceedings, 1985. Edited by L. Budach. XII, 533 pages. 1985.

Vol. 200: J. L. A. van de Snepscheut, Trace Theory and VLSI Design. VI, 0–140 pages. 1985.

Vol. 201: Functional Programming Languages and Computer Architecture. Proceedings, 1985. Edited by J.-P. Jouannaud. VI, 413 pages. 1985.

Vol. 202: Rewriting Techniques and Applications. Edited by J.-P. Jouannaud. VI, 441 pages. 1985.

Vol. 203: EUROCAL '85. Proceedings Vol. 1, 1985. Edited by B. Buchberger. V, 233 pages. 1985.

Vol. 204: EUROCAL '85. Proceedings Vol. 2, 1985. Edited by B. F. Caviness. XVI, 650 pages. 1985.

Vol. 205: P. Klint, A Study in String Processing Languages. VIII, 165 pages. 1985.

Vol. 206: Foundations of Software Technology and Theoretical Computer Science. Proceedings, 1985. Edited by S. N. Maheshwari. IX, 522 pages. 1985.

Vol. 207: The Analysis of Concurrent Systems. Proceedings, 1983. Edited by B. T. Denvir, W. T. Harwood, M. I. Jackson and M. J. Wray. VII, 398 pages. 1985.

Vol. 208: Computation Theory. Proceedings, 1984. Edited by A. Skowron. VII, 397 pages. 1985.

Vol. 209: Advances in Cryptology. Proceedings of EUROCRYPT '84. Edited by T. Beth, N. Cot and I. Ingemarsson. VII, 491 pages. 1985.

Vol. 210: STACS 86. Proceedings, 1986. Edited by B. Monien and G. Vidal-Naquet. IX, 368 pages. 1986.

Vol. 211: U. Schöning, Complexity and Structure. V, 99 pages. 1986.

Vol. 212: Interval Mathematics 1985. Proceedings, 1985. Edited by K. Nickel. VI, 227 pages. 1986.

Vol. 213: ESOP 86. Proceedings, 1986. Edited by B. Robinet and R. Wilhelm. VI, 374 pages. 1986.

Vol. 214: CAAP '86. 11th Colloquium on Trees in Algebra and Programming. Proceedings, 1986. Edited by P. Franchi-Zannettacci. VI, 306 pages. 1986.

Vol. 215: Mathematical Methods of Specification and Synthesis of Software Systems '85. Proceedings, 1985. Edited by W. Bibel and K. P. Jantke. 245 pages. 1986.

Vol. 216: Ch. Fernstrom, I. Kruzela, B. Svensson, LUCAS Associative Array Processor: Design, Programming and Application Studies. XII, 323 pages. 1986.

Vol. 217: Programs as Data Objects. Proceedings, 1985. Edited by H. Ganzinger and N. D. Jones. X, 324 pages. 1986.

Vol. 218: Advances in Cryptology – CRYPTO '85. Proceedings, 1985. Edited by H. C. Williams. X, 548 pages. 1986.

Vol. 219: Advances in Cryptology – EUROCRYPT '85. Proceedings, 1985. Edited by F. Pichler. IX, 281 pages. 1986.

Vol. 220: RIMS Symposia on Software Science and Engineering II. Proceedings, 1983 and 1984. Edited by E. Goto, K. Araki and T. Yuasa. XI, 323 pages. 1986.

Vol. 221: Logic Programming '85. Proceedings, 1985. Edited by E. Wada. IX, 311 pages. 1986.

Vol. 222: Advances in Petri Nets 1985. Edited by G. Rozenberg. VI, 498 pages. 1986.

Vol. 223: Structure in Complexity Theory. Proceedings, 1986. Edited by A. L. Selman. VI, 401 pages. 1986.

Vol. 224: Current Trends in Concurrency. Overviews and Tutorials. Edited by J. W. de Bakker, W.-P. de Roever and G. Rozenberg. XII, 716 pages. 1986.

Vol. 225: Third International Conference on Logic Programming. Proceedings, 1986. Edited by E. Shapiro. IX, 720 pages. 1986.

Vol. 226: Automata, Languages and Programming. Proceedings, 1986. Edited by L. Kott. IX, 474 pages. 1986.

Vol. 227: VLSI Algorithms and Architectures – AWOC 86. Proceedings, 1986. Edited by F. Makedon, K. Mehlhorn, T. Papatheodorou and P. Spirakis. VIII, 328 pages. 1986.

Vol. 228: Applied Algebra, Algorithmics and Error-Correcting Codes. AAECC-2. Proceedings, 1984. Edited by A. Poli. VI, 265 pages. 1986.

Vol. 229: Algebraic Algorithms and Error-Correcting Codes. AAECC-3. Proceedings, 1985. Edited by J. Calmet. VII, 416 pages. 1986.

Vol. 230: 8th International Conference on Automated Deduction. Proceedings, 1986. Edited by J. H. Siekmann. X, 708 pages. 1986.

Vol. 231: R. Hausser, NEWCAT: Parsing Natural Language Using Left-Associative Grammar. II, 540 pages. 1986.

Vol. 232: Fundamentals of Artificial Intelligence. Edited by W. Bibel and Ph. Jorrand. VII, 313 pages. 1986.

Vol. 233: Mathematical Foundations of Computer Science 1986. Proceedings, 1986. Edited by J. Gruska, B. Rovan and J. Wiedermann. IX, 650 pages. 1986.

Vol. 234: Concepts in User Interfaces: A Reference Model for Command and Response Languages. By Members of IFIP Working Group 2.7. Edited by D. Beech. X, 116 pages. 1986.

Vol. 235: Accurate Scientific Computations. Proceedings, 1985. Edited by W. L. Miranker and R. A. Toupin. XIII, 205 pages. 1986.

Vol. 236: TEX for Scientific Documentation. Proceedings, 1986. Edited by J. Désarménien. VI, 204 pages. 1986.

Vol. 237: CONPAR 86. Proceedings, 1986. Edited by W. Händler, D. Haupt, R. Jeltsch, W. Juling and O. Lange. X, 418 pages. 1986.

Vol. 238: L. Naish, Negation and Control in Prolog. IX, 119 pages. 1986.

Vol. 239: Mathematical Foundations of Programming Semantics. Proceedings, 1985. Edited by A. Melton. VI, 395 pages. 1986.

Vol. 240: Category Theory and Computer Programming. Proceedings, 1985. Edited by D. Pitt, S. Abramsky, A. Poigné and D. Rydeheard. VII, 519 pages. 1986.

Vol. 241: Foundations of Software Technology and Theoretical Computer Science. Proceedings, 1986. Edited by K.V. Nori. XII, 519 pages. 1986.

Vol. 242: Combinators and Functional Programming Languages. Proceedings, 1985. Edited by G. Cousineau, P.-L. Curien and B. Robinet. V, 208 pages. 1986.

Vol. 243: ICDT '86. Proceedings, 1986. Edited by G. Ausiello and P. Atzeni. VI, 444 pages. 1986.

Vol. 244: Advanced Programming Environments. Proceedings, 1986. Edited by R. Conradi, T.M. Didriksen and D.H. Wanvik. VII, 604 pages. 1986

Vol. 245: H.F. de Groote, Lectures on the Complexity of Bilinear Problems. V, 135 pages. 1987.

Vol. 246: Graph-Theoretic Concepts in Computer Science. Proceedings, 1986. Edited by G. Tinhofer and G. Schmidt. VII, 307 pages. 1987.

Vol. 247: STACS 87. Proceedings, 1987. Edited by F.J. Brandenburg, G. Vidal-Naquet and M. Wirsing. X, 484 pages. 1987.

Vol. 248: Networking in Open Systems. Proceedings, 1986. Edited by G. Müller and R.P. Blanc. VI, 441 pages. 1987.

Vol. 249: TAPSOFT '87. Volume 1. Proceedings, 1987. Edited by H. Ehrig, R. Kowalski, G. Levi and U. Montanari. XIV, 289 pages. 1987.

Vol. 250: TAPSOFT '87. Volume 2. Proceedings, 1987. Edited by H. Ehrig, R. Kowalski, G. Levi and U. Montanari. XIV, 336 pages. 1987.

Vol. 251: V. Akman, Unobstructed Shortest Paths in Polyhedral Environments. VII, 103 pages. 1987.

Vol. 252: VDM '87. VDM – A Formal Method at Work. Proceedings, 1987. Edited by D. Bjørner, C.B. Jones, M. Mac an Airchinnigh and E.J. Neuhold. IX, 422 pages. 1987.

Vol. 253: J.D. Becker, I. Eisele (Eds.), WOPPLOT 86. Parallel Processing: Logic, Organization, and Technology. Proceedings, 1986. V, 226 pages. 1987.

Vol. 254: Petri Nets: Central Models and Their Properties. Advances in Petri Nets 1986, Part I. Proceedings, 1986. Edited by W. Brauer, W. Reisig and G. Rozenberg. X, 480 pages. 1987.

Vol. 255: Petri Nets: Applications and Relationships to Other Models of Concurrency. Advances in Petri Nets 1986, Part II. Proceedings, 1986. Edited by W. Brauer, W. Reisig and G. Rozenberg. X, 516 pages. 1987.

Vol. 256: Rewriting Techniques and Applications. Proceedings, 1987. Edited by P. Lescanne. VI, 285 pages. 1987.

Vol. 257: Database Machine Performance: Modeling Methodologies and Evaluation Strategies. Edited by F. Cesarini and S. Salza. X, 250 pages. 1987.

Vol. 258: PARLE, Parallel Architectures and Languages Europe. Volume I. Proceedings, 1987. Edited by J.W. de Bakker, A.J. Nijman and P.C. Treleaven. XII, 480 pages. 1987.

Vol. 259: PARLE, Parallel Architectures and Languages Europe. Volume II. Proceedings, 1987. Edited by J.W. de Bakker, A.J. Nijman and P.C. Treleaven. XII, 464 pages. 1987.

Vol. 260: D.C. Luckham, F.W. von Henke, B. Krieg-Brückner, O. Owe, ANNA, A Language for Annotating Ada Programs. V, 143 pages. 1987.

Vol. 261: J. Ch. Freytag, Translating Relational Queries into Iterative Programs. XI, 131 pages. 1987.

Vol. 262: A. Burns, A.M. Lister, A.J. Wellings, A Review of Ada Tasking. VIII, 141 pages. 1987.

Vol. 263: A.M. Odlyzko (Ed.), Advances in Cryptology – CRYPTO '86. Proceedings. XI, 489 pages. 1987.

Vol. 264: E. Wada (Ed.), Logic Programming '86. Proceedings, 1986. VI, 179 pages. 1987.

Vol. 265: K.P. Jantke (Ed.), Analogical and Inductive Inference. Proceedings, 1986. VI, 227 pages. 1987.

Vol. 266: G. Rozenberg (Ed.), Advances in Petri Nets 1987. VI, 451 pages. 1987.

Vol. 267: Th. Ottmann (Ed.), Automata, Languages and Programming. Proceedings, 1987. X, 565 pages. 1987.

Vol. 268: P.M. Pardalos, J.B. Rosen, Constrained Global Optimization; Algorithms and Applications. VII, 143 pages. 1987.

Vol. 269: A. Albrecht, H. Jung, K. Mehlhorn (Eds.), Parallel Algorithms and Architectures. Proceedings, 1987. Approx. 205 pages. 1987.

Vol. 270: E. Börger (Ed.), Computation Theory and Logic. IX, 442 pages. 1987.

Vol. 271: D. Snyers, A. Thayse, From Logic Design to Logic Programming. IV, 125 pages. 1987.

Vol. 272: P. Treleaven, M. Vanneschi (Eds.), Future Parallel Computers. Proceedings, 1986. V, 492 pages. 1987.

Vol. 273: J.S. Royer, A Connotational Theory of Program Structure. V, 186 pages. 1987.

Vol. 274: G. Kahn (Ed.), Functional Programming Languages and Computer Architecture. Proceedings. VI, 470 pages. 1987.

Vol. 275: A.N. Habermann, U. Montanari (Eds.), System Development and Ada. Proceedings, 1986. V, 305 pages. 1987.

Vol. 276: J. Bézivin, J.-M. Hullot, P. Cointe, H. Lieberman (Eds.), ECOOP '87. European Conference on Object-Oriented Programming. Proceedings. VI, 273 pages. 1987.

Vol. 277: B. Benninghofen, S. Kemmerich, M.M. Richter, Systems of Reductions. X, 265 pages. 1987.

Vol. 279: J.H. Fasel, R.M. Keller (Eds.), Graph Reduction. Proceedings, 1986. XVI, 450 pages. 1987.

Vol. 280: M. Venturini Zilli (Ed.), Mathematical Models for the Semantics of Parallelism. Proceedings, 1986. V, 231 pages. 1987.

Vol. 281: A. Kelemenová, J. Kelemen (Eds.), Trends, Techniques, and Problems in Theoretical Computer Science. Proceedings, 1986. VI, 213 pages. 1987.

Vol. 282: P. Gorny, M.J. Tauber (Eds.), Visualization in Programming. Proceedings, 1986. VII, 210 pages. 1987.

Vol. 283: D.H. Pitt, A. Poigné, D.E. Rydeheard (Eds.), Category Theory and Computer Science. Proceedings, 1987. V, 300 pages. 1987.

Vol. 284: A. Kündig, R.E. Bührer, J. Dähler (Eds.), Embedded Systems. Proceedings, 1986. V, 207 pages. 1987.

Vol. 285: C. Delgado Kloos, Semantics of Digital Circuits. IX, 124 pages. 1987.

Vol. 286: B. Bouchon, R.R. Yager (Eds.), Uncertainty in Knowledge-Based Systems. Proceedings, 1986. VII, 405 pages. 1987.

Vol. 287: K.V. Nori (Ed.), Foundations of Software Technology and Theoretical Computer Science. Proceedings, 1987. IX, 540 pages. 1987.

Vol. 288: A. Blikle, MetaSoft Primer. XIII, 140 pages. 1987.